Requirements Engineering and Management for Software Development Projects

T0213924

Murali Chemuturi

Requirements Engineering and Management for Software Development Projects

Foreword by Tom Gilb

 Springer

Murali Chemuturi
Chemuturi Consultants
murali@chemuturi.com

ISBN 978-1-4899-9307-6   ISBN 978-1-4614-5377-2 (eBook)
DOI 10.1007/978-1-4614-5377-2
Springer New York Heidelberg Dordrecht London

Printed on acid-free paper

Springer is part of Springer Science+Business Media (www.springer.com)

Foreword

This book is a practical textbook which will be useful for a requirements student and/or a software manager student, to get a picture of the very many practical considerations that go into specifying and validating requirement for IT/IS projects.

The book does not oversimplify subjects that require mature consideration in order to succeed. In my view the author gets many and most critical points correct, better than the many less mature authors.

For a small example of such points, the importance of stakeholders, the silliness of the *non-functional requirement* term, and an understanding that quality is designed in, not tested in.

The pages are, like my own detailed work, dense with powerful and useful lists of considerations. They will give excellent structure to a teacher who can help students discuss the points, and explain to students using examples.

I hope this textbook finds its place as a teaching tool for information technology courses. But the lone reader can safely use it as a mature way to survey the entire software development scene today.

Kolbotn, Norway, 8 June 2012 Tom Gilb

Preface

Gerald M. Weinberg, author of the book "The Psychology of Computer Programming" is attributed with the quote —"*If builders built houses the way programmers built programs, the first woodpecker to come along would destroy civilization.*" I could not agree more. The rate of project failure is much higher in software development compared to either manufacturing or construction. It is not that there are no failures in manufacturing or construction. Those failures are in "first-of-its-kind" projects, especially in manufacturing. In construction, these are even fewer. For example, the Empire State Building of New York city was the first of its kind when it was built. It is the first building in the world to go up 80 floors high above ground. It was the tallest building in the world for a number of years. The issues, there would have been many, were solved in the specifications and the design stage. The construction would scrupulously adhere to the design. It was a success.

Why do software projects fail at such a high rate even when there were similar projects executed earlier?

Two major causes are attributed for this phenomenon. The first is the poor understanding and definition of product requirements. This leads to technical failure. The second is the poor project management of developing the software product to the specified requirements resulting in unsustainable overruns of cost and schedule. Both the reasons lead to project failure.

In this book, I am focusing on the precise understanding and definition of product requirements. As in other areas, there is more misunderstanding about this critical activity than right understanding.

"*The hardest single part of building a software system is deciding what to build... No other part of the work so cripples the resulting system if done wrong. No other part is more difficult to rectify later.*" said Frederick Brooks, Jr., Brooks Computer Science Building, University of North Carolina, USA. (From his paper "No Silver Bullet: Essence and Accident in Software Engineering," 1986 as also *The Mythical Man-Month: Essays on Software Engineering,* Anniversary Edition, Chap. 16.) I concur wholeheartedly.

Unfortunately for me or perhaps for the software development industry itself, there is no commonly accepted taxonomy for software engineering activities. It is not that there are no definitions at all. The definitions that are there, are not universally accepted. Some, like "non-functional requirements" are downright ridiculous. That is the reason why I am explaining every term I use here, as you will notice, from the first fundamentals. Wherever, possible, I am using the terminology from a credible source. I am giving all the available meanings with my own interpretation with the idea that the reader is better informed when the same matter is presented by someone else with a different set of nomenclature. Please bear with me for this over-specification.

My own perception was gathered from my experience, observation, study and participation in discussion forums of how well the requirements are either engineered or managed. It is not very flattering to the community of requirements engineers or managers. There are many doubts, which begin right at the fundamental stage among the practitioners. This is, perhaps, due to the fact that universities are not conducting courses in requirements engineering and management. Their focus is more on engineering and producing code rather than on the forward or backward linkages to code production. Managements rather hasten the project teams into coding ASAP. Of course, there are exceptions without which, I would not have been able to gather best practices.

In this book, I tried to give you a complete view of the activities of requirements engineering as well as requirements management. Both the activities, engineering and management, are equally important. Engineering activities, performed well, produce the right deliverable. When we manage the engineering activities diligently, we produce the deliverable within the estimated cost and on schedule. The variances that are bound to be there would be predictable and within acceptable levels. Management activities when performed diligently would also allow us to plow the experience back into the process of performing engineering activities and facilitate improvement.

The information presented here is from my experience, observation, academic study and participation in the Internet discussion forums.

That was my intent and I would like to learn how you perceived my effort to be. Please feel free to email me at murali@chemuturi.com and I promise to respond to every email that I receive normally in one business day.

June 2012   Murali Chemuturi

Acknowledgments

When I look back, I find that there are so many people to whom I should be grateful. Be it because of their commissions or omissions, they made me a stronger and a better person, and both directly and indirectly helped to make this book possible. It would be difficult to acknowledge everyone's contributions here, so to those whose names may not appear, I wish to thank you all just the same. I will have failed in my duty if I did not explicitly and gratefully acknowledge the persons below:

- My parents, Appa Rao and Vijaya Lakshmi, the reason for my existence. Especially my father, a rustic agrarian, who by personal example taught me the virtue of hard work and how sweet the aroma of sweat from the brow can be.
- My family, who stood by me like a rock in difficult times. Especially my wife, Udaya Sundari, who gave me the confidence and the belief that "I can." And my two sons, Dr. Nagendra and Vijay, who provided me the motive to excel.
- My two uncles, Raju and Ramana, who by personal example taught me what integrity and excellence mean.
- Springer Science+Business Media and especially Ms. Susan Lagerstrom-Fife and Ms. Courtney Clark for their belief in the content of this book, for their generous allocation of time, and for leading me by the hand like the Good Lord through every step of making this book a reality.
- The staff of Springer Science+Business Media all of whom were involved in bringing this book out to the public.

To all of you, I humbly bow my head in respect, and salute you in acknowledgement of your contribution.

Murali Chemuturi

Contents

Abbreviations

ANSI	American National Standards Institute
ASD	Adaptive Software Development
AUP	Agile Unified Process
BFS	Business Function Specification
CCB	Change/Configuration Control Board
CIO	Chief Information Officer
CMMI	Capability Maturity Model Integration
COBOL	Common Business Oriented Language
COTS	Commercial Off The Shelf
CPU	Central Processing Unit
CR	Change Request
CRM	Customer Relationship Management
CRR	Change Request Register
DBA	Database Administrator
DBMS	Database Management System
DDS	Detail Design Specification
DFD	Data Flow Diagram
DIR	Defect Injection Rate
DSDM	Dynamic Systems Development Method
EAI	Enterprise Architecture Integration
EDI	Electronic Data Interchange
EDP	Electronic Data Processing
ERD	Entiry-Relationship Diagram
ERP	Enterprise Resources Planning
FDD	Feature Driven Development
FDS	Functional Design Specification
FS	Functional Specification
GMP	Good Manufacturing Practice
GPM	Gross Productivity Metric for Requirements Engineering
GSDP	Good Software Development Practice
GUI	Graphical User Interface

HLD	High Level Design
HR	Human Resources
IDE	Interactive Development Environmet
IEEE	Institute of Electrical and Electronics Engineers
ISO	International Organization for Standardization
IT	Information Technology
IV&V	Independent Verification and Validation
JSS	Joint Services Specification
LLD	Low Level Design
LOC	Lines of Code
MBA	Master of Business Administration
MIS	Management Information System
NASSCOM	National Association of Software and Services Companies of India
NIU	Network Interface Unit
OOAD	Object Oriented Analysis and Design
OOM	Object Oriented Methodology
PEG	Productivity for Elicitation and Gathering
PER	Productivity for Establishing the Requirements
PM	Project Management
RAM	Random Access Memory
RCL	Requirements Capture Language
REC	Relative Effort spent on resolving CRs
RECC	Relative Effort spent on a specific Change Request Category
REM	Requirements Engineering and Management
RM	Requirements Management
RQC	Relative Effort metric for Quality Control of Requirements engineering activities
ROI	Return On Investment
RSM	Requirements Stability Metric
RUP	Rational Unified Process
SCM	Supply Chain Management
SCMP	Software Configuration Management Plan
SDLC	Software Development Life Cycle
SDS	Software Design Specification
SMRE	Schedule Metric for Requirements Engineering
SOP	Standard Operating Procedure
SOW	Statement of Work
SPIN	Software Process Improvement Network
SPMN	Software Project Managers Network
SPMP	Software Project Management Plan
SQAP	Software Quality Assurance Plan
SRS	Software Requirements Specification
SSADM	Structured Systems Analysis and Design Method
SyRS	System Requirements Specification

TCM	Test Coverage Metric
TDD	Test Driven Development
TQM	Total Quality Management
UI	User Interface
UML	Unified Modeling Language
URS	User Requirements Specification
VARS	Value Added Re-Sellers
WIP	Work in Process
XP	Extreme Programming
Y2K	Year 2000

Chapter 1
Introduction to Requirements Engineering and Management

Requirements are the precursor to all other software development phases, namely, software design, software construction and testing. When the end result of the software development activity is a COTS (Commercial Off The Shelf) product, we term requirements as "product specifications". When the end result of the software development activity is to deliver the product to a single client in a project scenario, we use the term requirements. In either case, the activity of managing the requirements is the same.

The importance of properly managing requirements cannot be overemphasized as any omission of a vital requirement or error committed during requirements analysis results in increased cost of the product and in some cases, may result in project / product failure. Another important aspect of requirements management is the change management. If changes to requirements are not properly controlled, it may result in uncontrollable scope creep and increased costs.

A proper understanding of requirements and careful management thereof can prevent project failures and contribute to the delivery of quality software products to intended clients.

1.1 What is a "Requirement"

Consider these five statements:

1. I "hope" to have a car (The capability to posses a car is absent but hope exists that someday it might be possible)
2. I "wish" to have a car (The capability to posses a car is distinctly possible but not feasible yet)
3. I "desire" to have a car (The capability to posses a car exists. But there are other competing demands to cater to.)

M. Chemuturi, *Requirements Engineering and Management for Software Development Projects*, DOI: 10.1007/978-1-4614-5377-2_1,
© Springer Science+Business Media New York 2013

4. I "need" a car (The capability exists and it is feasible. Having a car surpassed other competing demands)
5. I "require" a car (Possessing a car can no longer be postponed. It is essential now)

Could you see the increasing emphasis with each statement as it moves from statement 1 to statement 5? The term "Requirement" connotes essentiality of some need.

The dictionary defines requirements as "a need", "a thing needed", "a necessary condition", "a demand", "something essential to the existence or occurrence of something else", and "something that is needed or that must be done".

Simply stated, a requirement is a need of some person or process. A requirement is capable of being fulfilled. If we come across a requirement that cannot be fulfilled, it becomes a desire that can perhaps be fulfilled at a later date or with a better technology or better set of circumstances.

Wikipedia defines requirements in the context of software engineering thus, "*It is a statement that identifies a necessary attribute, capability, characteristic, or quality of a system in order for it to have value and utility to a user*". In the context of other engineering disciplines, it defines requirements as "*a singular documented need of what a particular product or service should be or perform*".

IEEE (Institute of Electrical and Electronics Engineers) standard 610 "Glossary of Software Engineering Terminology" provides three definitions:

1. A condition or capability needed by a user to solve a problem or achieve an objective,
2. A condition or capability that must be met/possessed by a system or system component to satisfy a contract, standard, specification or other formally imposed documents.
3. A documented representation of a condition or a capability as in (1) or (2) above.

CMMI® (Capability Maturity Model Integration) for Development version 1.3 also gives three definitions almost similar to IEEE definitions:

1. A condition or capability needed by a user to solve a problem or achieve an objective,
2. A condition or capability that must be met/possessed by a product, service, product component or service component to satisfy a supplier agreement, standard, specification or other formally imposed documents.
3. A documented representation of a condition or a capability as in (1) or (2) above.

As can be seen, CMMI definitions are slightly different that too, only in the second definition and that difference is only minimal. It appears that CMMI adopted the IEEE definitions.

The above definitions suffer from the following limitations:

1. They talk only about needs – that is, only the essential aspects. The do not take into consideration, the reasonable expectations of the users which expect the development team to bring their expertise to bear on the software product
2. They do not take into consideration the constraints of the users with which they have to live with, even if the software offers better options. For example there are many situations where authentication by the signature of the hand is essential in spite of the great advances in the field of digital signatures.
3. They do not take into consideration the interfaces with other and perhaps existing systems. This is especially so in cases where a COTS product like ERP (Enterprise Resources Planning) or CRM (Customer Relationship Management) type of product implementations.
4. They do not place the responsibility for the requirements. By not including the stakeholders who can state needs, the developers are mislead and miss out on some of the essential stakeholders while defining the requirements.
5. It ignores unstated needs. It gives the connotation that "if it is not documented, it is not considered". This is perhaps, the origin of the joke, especially in the matter of Microsoft's Windows operating system, that it has a Start button but no Stop button!

The software development industry is generally adhering to these definitions. But circumstances are continuously changing especially so in the field of software and its development. We need a more comprehensive definition. Summarizing the above discussion, we may define "Requirements" in the context of software development projects, in a more comprehensive manner, thus,

A requirement is a need, expectation, constraint or interface of any stakeholders that must be fulfilled by the proposed software product during its development".

This definition has the following key terms:

1. **Need**—It is something basic without which the existence becomes untenable. It is the absolute minimum necessity if the system is to be useful. If a need is not met, the system becomes unusable or less usable.
2. **Expectation**—Expectation is an unstated need. When users entrust the development of software to a team (in-house or outsourced) it is expected that the development team brings expertise of software to bridge the gap in the needs stated by the user.
3. **Constraint**—It is a hurdle that the user has to live with. It may be in terms of a limitation on the leverage of the software design or development.
4. **Interface**—It is the basis for interaction with the customers, suppliers, and peers (in the forward chain or backward chain) of the user.
5. **Stakeholders**—A stakeholder is someone who is affected by the outcome of a human endeavor. A software development project has multiple stakeholders, namely,

   a. The end user who is the ultimate user of the product
   b. The project team that is going to develop the product to fulfill the need

    c. The marketing team, if the resultant product is a COTS product so that they can find customers and sell it

    d. The managements of both the supplier and the customer as both derive ROI (Return on Investment) from the endeavor

6. **That must be fulfilled**—The need must be fulfilled. If it cannot be fulfilled either due to limitations of technology or finance, it becomes a future requirement. If the need cannot be fulfilled by the present endeavor, then the endeavor itself becomes unnecessary

7. **The proposed software product**—It is the place where the need is expected to be fulfilled. It is the end result of the present endeavor

8. **During its development**—This specifies the timeline when the need shall be fulfilled. If it is not being fulfilled during present development, then the need remains unfulfilled or a future need.

When we come to the software development arena, we have two types of requirements namely the user requirements and the software requirements.

User requirements are the needs specified by the end user for the proposed software product. They consist mainly of functionalities to be achieved by the software and any conveniences that are needed in improving the personal performance. Users may also require additional analyses that may be not possible in the present system but are necessary in improving the performance of the system or to serve the customers in a better manner.

Software requirements are additions to the user requirements that are unique to software systems like usability, security, user friendliness, audit trails and so on.

Software Requirements is a term used by IEEE. Even then, IEEE did not define this phrase in its glossary of software engineering terminology standard in spite of defining a standard for "Software Requirements Specifications". Even the CMMI for development version 1.3 also does not mention this phrase.

1.2 Requirements Management

Having put the term "requirement" in its proper perspective, let us now look at the phrase "Requirements Management", the topic that is covered by this book. As can be readily seen the two words in this phrase are "Requirements" and "Management". We have already defined and put the term "Requirements" in its proper perspective.

Now let us look at the term "Management". This term has three connotations.

1. The first connotation is as a "group of individuals" who are running the affairs of the company. They normally comprise of the Board of Directors and other senior executives designated by the Board as Managers.

2. The second connotation is as an "art and a social science (process)" practiced by individuals to get things done by others without absolute authority or clarity in the organizational processes.
3. The third connotation is as a "body of knowledge" about getting things done

It is the second connotation that is pertinent to the present context of requirements management. Management as a process consists of the following sub processes:

1. Planning
2. Organizing
3. Staffing
4. Controlling

Therefore, requirements management includes the above four aspects. CMMI Version 1.3 defines requirements management, as, *"The management of all requirements received by or generated by the project or work group, including both technical and non-technical requirements levied on the project or work group by the organization"*. Technical requirements are defined as *"properties of product or service to be acquired or developed"* and non-technical requirements are defined as *"requirements affecting product and service acquisition or development that are not properties of the product or service"*.

This definition specifies that requirements can be of two varieties, namely technical and non-technical requirements. It also states that requirements can be received from external sources such as customers or can be generated within the project team too. Especially in COTS (Commercial Off The Shelf) product scenario, many, if not all, requirements are generated within the project team.

Wikipedia defines requirements management, as, *"Requirements management is the process of documenting, analyzing, tracing, prioritizing, and agreeing on requirements and then controlling and communicating to relevant stakeholders. It is a continuous process throughout the project"*.

This definition enumerates the activities to be performed as part of requirements management, namely,

1. Documenting requirements
2. Analyzing the requirements
3. Tracing the requirements through out the development life cycle
4. Prioritizing the requirements, especially their order of implementation
5. Agreeing upon the requirements, that is requirements are accepted for implementation by the project team and are approved by the stakeholders
6. Controlling the requirements, that is controlling the change to the agreed upon requirements
7. Communicating the status and progress of implementation of requirements and changes received thereon to all stakeholders.

This definition also states that the process of requirements management begins with the starting of the projects and completes with the ending of the project.

Both are good definitions and cover requirements management in an apt manner and help us in understanding the subject fully. While the CMMI definition covers the activities using the term "management", the Wikipedia definition enumerates the activities. The Wikipedia definition can be considered as the continuation of the CMMI definition of requirements management.

As part of management, we plan for ensuring that right requirements (that are complete, exhaustive and clear) are made available to the development team. The plan would include activities:

1. for collecting, analyzing and establishing the project requirements
2. for ensuring that changes to the established requirements are carried out in a controlled manner
3. for ensuring that all requirements are traced through the development life cycle and are delivered to the customer effectively.

The organizing part consists of creating and maintaining an environment that is conducive to carry out requirements related activities efficiently and effectively. It includes defining processes for carrying out the required activities as well as ensuring that quality is built into the deliverables. It also includes defining processes for ensuring that changes to requirements are made in a controlled manner.

Staffing process includes recruiting qualified personnel to carry out requirements related activities; providing them necessary training; providing necessary tools and techniques; and keep them motivated.

Controlling is ensuring that all the above three activities are carried out conforming to the corresponding plans and making mid-course corrections as and when required to ensure that deliveries are made on time and with the best attainable quality.

1.3 Requirements Management Scenarios

Requirements need to be managed during software development which is carried out for the following purposes:

1. When an organization wishes to shift a set business process from manual processing system to a computer-based processing system

   a. When the requisite software is developed using its internal software development department, the project is referred to as "in-house project".
   b. When the requisite software is outsourced to an external software development organization, the project is referred to as "external project".

2. When an organization decides to shift an older computer-based system (perhaps a batch processing system) to a better computer-based system (perhaps a web-based processing system),

a. When the requisite software is developed using its internal software development department, the project is referred to as "upgrade in-house project",

b. When the requisite software is outsourced to an external software development organization, the project is referred to as "upgrade external project"

3. When an organization decides to develop a software product for selling to various customers, which I call "product development".
4. When an organization decides to overhaul their existing software product and upgrade it to next level which I call "product upgrade"

Scenarios 1 and 2, are commonly referred to as "project development" and scenarios 3 and 4 are commonly referred to as "product development". I will be using these two terms in this book.

1.4 Agencies Responsible for Managing Requirements

Basically it is the project manager who has to manage the requirements for the project, but others do have a role in managing the requirements. In the project development scenario, two agencies need to manage requirements:

1. In the in-house project development scenario, the software project manager is responsible for managing the project requirements. Of course, he can delegate this activity to a business analyst on the team.
2. In the external project scenario, two persons are responsible for managing the requirements:

a. The project coordinator at the outsourcing organization
b. The software project manager at the outsourced organization

3. In the product development scenario

a. Product Manger who is normally from the marketing department and is usually in-charge of selling the final product manages the requirements in the organization
b. The project manager who is leading the software development team.

The above mentioned agencies are primarily responsible for managing software project's requirements. It does not mean that other stakeholders are free of any responsibility in respect of project requirements management. The rest of the stakeholders have the secondary responsibility.

The senior management of the organization have the responsibility of providing resources for the activity and to ensure that the activity is being carried out diligently. The end-users are responsible to provide the requirements comprehensively and lucidly as well as to provide clarifications whenever needed by the project team. The business analysts are vested with the responsibility to accurately record the requirements and convey them to the development team. The development team has the responsibility to build and deliver the software conforming to the requirements as well as to ensure that all accepted requirements are fully met.

1.5 Approaches to Requirements Management

There are two schools or thought on this aspect. One school of thought states that the project requirements must be managed methodically and diligently conforming to a defined process which is continuously improved in the organization. The other school of thought is that there is no such need for expending special effort for managing project requirements as it is a natural part of software development.

The salient aspects of the first school of thought are:

1. The organization needs to be a process-driven organization. That is, the organization must have a defined process; the process is diligently implemented within the organization; the process must be internalized; the organizational process is continuously improved conforming to a defined process for improvement.
2. The organizational process would have a set of procedures, standards, guidelines, formats and templates for managing requirements.
3. The effectiveness of the implemented process is measured and corrective action is taken to correct / improve the process as necessary.
4. Deviations from the defined process are allowed and waivers given based on the specific set of conditions conforming to the *tailoring guidelines* specified in the process.

The main argument in favor of a process-driven approach is that uniformity can be achieved across the organization in the matter of requirements management. This is a great advantage for software development organizations executing multiple projects concurrently. If the requirements management is allowed without any controls, it is likely to derail the projects and result in failed project execution. The other advantages are that new project managers can perform on par with the experienced ones; experienced ones can perform at a higher level of performance; and it provides predictability for performance for everyone concerned in the project execution. Process-driven approach places the onus for performance and results on the process than on the individual.

The proponents of ad-hoc management for project requirements argue that there is no inherent need for uniformity across projects in the organization; the process puts overhead and reduces the productivity of the project resources; the measure of ultimate success of the project is not in how well the requirements are managed but how well the end product performs; and finally, that the management of requirements has little impact on the final product. They also point out that however well the process may have been defined, if the individual implementing it is weak, the results would be disastrous. They say that it is better to invest in a capable individual and trust him/her to get the results than invest in a process. Ad-hoc approach places onus on the individual to obtain results. Heroics are possible. Adherents of agile methodologies prefer ad-hoc approach to requirements management in true adherence to agile philosophy which puts emphasis on customer satisfaction over everything else. Agile projects use iterative life cycle for software development in which the total software product is developed and delivered in iterations with no iteration being longer than four weeks.

It is my observation that the methodical approach gets results relatively more often than an ad-hoc approach. Imagine yourself going to a doctor to cure some malady. Let us say that the doctor does not believe in process-driven approach. Assume for a minute that the doctor tells you "Here, take this medicine. If it doesn't work come back and I will give you another medicine. I will cure your malady with an iterative approach. It is much faster and cheaper to cure your trouble than going through all those unreliable diagnostic procedures. What do you say?"

Are you likely to continue your treatment with that doctor? The chances are very slim—right? You would go to a doctor who first diagnoses your malady using a series of diagnostic procedures than jumping into treatment right away, even though the doctor costs you more money. But here we are in software development organizations, advocating jumping into coding right away and thumb our noses at methodical approach.

Do I sound like condemning ad-hoc approach completely?

Here are my recommendations:

1. When the organization is small with the owner managing the projects, ad-hoc approach would deliver results.
2. If the organization has few concurrent projects, say less than five, the ad-hoc approach would be adequate.
3. When the functional domain in which the organization operates is the same in all projects and the human resources are stable with little attrition, again the ad-hoc approach would get the results.
4. If the organization is large, handles multiple projects concurrently, has significant attrition in the human resources and the functional domains are different in every project, a process-driven approach is perhaps mandatory.

1.6 Requirements Engineering

The word "Engineering" has multiple meanings and connotations.

One definition of the term "engineering" is "*engineering is the application of science for practical purposes*". It connotes that all engineering is based on proven scientific principle and therefore *engineering is not art*.

Another definition of the term engineering is that *engineering is a field of study and research*. The fields of study and research such as mechanical engineering, electrical engineering, construction engineering, chemical engineering, electronics engineering and so on are all collectively known as engineering.

Engineering is also defined as a process. It *is a process of converting the specifications of customers into such artifacts that are used by artisans to produce the product that fulfills the customer specifications.* The artifacts can be engineering drawings, process documents, parts lists, material specifications and so on. The artifacts can even be textual documents.

The definition relevant to requirements engineering is the last one that it is a process. However, requirements engineering does not cover the entire definition. It only covers the definition partially. The activities included in requirements engineering are:

1. Collecting the requirements from customers
2. Compiling and collating the requirements
3. Establishment of the requirements
4. Ensuring the integrity of the requirements
5. Tracing, tracking and reporting the progress of requirements through the software development life cycle.

Each of these activities are discussed in detail in the following chapters.

1.7 Topics Proposed to be Covered in this Book

Having placed both "Requirements" and "Requirements Management" in their proper perspective, we are now ready to move forward. We will be learning all the above topics in greater detail in the subsequent chapters.

This book aims at consolidating my theoretical knowledge garnered from reading books; learning from the knowledgeable seniors; my own experience in software development; observation of peer projects; and my experience in conducting training programs to the learners of the discipline of requirements management. This consolidation - it is my fervent hope would aid the learners as a guide for learning and the experienced professionals in the field as a reference text.

The following topics are covered to be covered in this book:

1. Introduction to Requirements Management—covered in this chapter
2. Understanding Requirements
3. Elicitation / Gathering Requirements
4. Requirements Analysis and Development
5. Establishment of Requirements
6. Verification and validation of requirements
7. Planning for Requirements Management
8. Change Management in Requirements Management
9. Requirements, Tracing, Tracking and Progress Reporting
10. Metrics & Measurement in Requirements Management

11. Roles & Responsibilities in Requirements Management
12. RM through SDLC
13. Tools and techniques for use in Requirements Management

    13.1 SSADM
    13.2 OOAD
    13.3 UML
    13.4 Agile methods
    13.5 Any other organization specific tools

14. Pitfalls and best practices in Requirements Management.

Chapter 2
Understanding Requirements

We have defined the term "requirement" in Chap. 1 as applied in the context of requirements management in software development. Now let us discuss the requirements in greater detail so that we understand the term in its entirety.

2.1 Classification of Requirements

Requirements can be classified based on three considerations, namely,

1. Functionality considerations—these are the requirements that fulfill the set of selected business processes and deliver the results to end-users.
2. Product construction considerations—these are the requirements that are necessary to build the product efficiently as well as to maintain it later on.
3. Source considerations—requirements for software development are provided from different sources. This classification is based on the agencies that provide the requirements.

2.2 Classification of Requirements Based on Functionality Considerations

Requirements can be classified into two major classes from the functionality standpoint, namely,

1. Core functionality requirements
2. Ancillary functionality requirements

M. Chemuturi, *Requirements Engineering and Management for Software Development Projects*, DOI: 10.1007/978-1-4614-5377-2_2,
© Springer Science+Business Media New York 2013

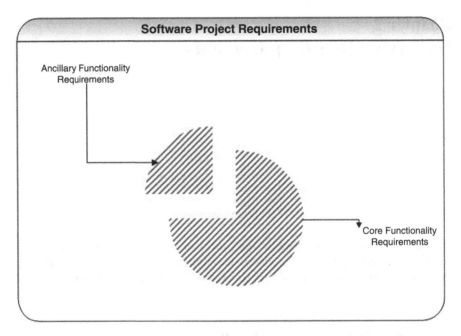

Fig. 2.1 Classification of software project requirements based on functionality

Core functionality requirements are those functionalities of the product without which, the product is not useful for the users. These functionalities must be fulfilled and exceptions cannot be granted or resorted to during development of the product. Core functionality addresses the performance of a set of business processes. The main purpose of software development is to fulfill this core functionality.

Ancillary functionality requirements supplement core functionality. Even if the ancillary functionality is not fulfilled, the product is still useful but may cause inconvenience in the form of loss of productivity or security. One significant point to be noted here is that the customer, more often than not, may not specify this functionality and even expects the development team to take care of this ancillary functionality!

Figure 2.1 depicts the classification based on functionality pictorially.

Classification of requirements based on functionality consideration is sometimes classified as (a) Functional Requirements and (b) Non-Functional Requirements. When we prefix the word "non" to any other word, it connotes the opposite of the word. When we say "non-functional" it has connotation that the requirements do not function or do not serve any function. In reality these requirements may not serve business process functions directly but they are serving a useful purpose in the software, They indirectly, perhaps, assist in the better functioning of business processes. User-friendliness does not serve any business process but if we take it away from our software, using the software and performing the business

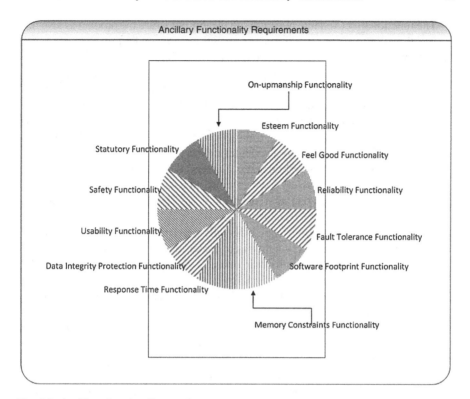

Fig. 2.2 Ancillary functionality requirements

processes using the software becomes tedious. That is the reason, I have not used the "functional and non-functional" classification of requirements.

In almost all cases of software development, core functionality is addressed completely. The customer provides core functionality in project scenario. It is defined by the project team or domain experts in the case of a product development scenario. But when it comes to ancillary functionality, it is common to miss out on some of the ancillary functionality. In many cases where the end product of software development is panned by the critics, the criticism stems from the fact that all relevant ancillary functionality is not fully or not efficiently implemented.

Various classes of statutory functionality requirements is depicted pictorially in Fig. 2.2.

The following are the ancillary functionalities that need to be included in the overall requirements for the project:

1. **Statutory functionality**—This functionality is required because of regulations/ standards of government, or industry associations or professional associations. In software development industry we have Institute of Electrical and Electronics Engineers (IEEE), Software Process Improvement Networks (SPINs), Software Project Managers Networks (SPMNs) in addition to governments.

The functionality to enable persons with disabilities to use the product as also the functionality to ensure that intruder prevention come under this category.

2. **Safety functionality**—This functionality protects the user from causing injury to the users. However, as computer usage especially from this stand point cannot cause physical injury, it has to protect him from logical injuries including financial losses. Examples for safety functionality include charging the credit card but not booking a ticket in online transactions, computational inaccuracies, performing wrong transaction (buying shares when we intended to sell), and data loss due to accidental deletes.

3. **Security functionality**—While safety functionality is to protect the user from the product, security is safety from external attacks. Protection from intruders, malicious use of insiders, and data theft via the Internet and so on are examples of security functionality.

4. **Usability functionality**—It was originally referred to as User-friendliness. The objective of this functionality is to make the software usable intuitively and with minimal reference to user manuals. Graphical user Interface (GUI) has achieved much of this functionality but improvement is always possible.

5. **Data Integrity Protection functionality**—Now that computing shifted from EDP (Electronic Data Processing) rooms to end users, this has become an important functionality for any software product. End user can unintentionally and innocently affect data integrity. They may enter numerals in name fields and alphabets in numeric fields. In Human Resources (HR) applications, they may enter a date of birth such that the age of the employee may either be below legal employment age or above the retirement age. In hospital or hotel management applications, they may enter check out date as prior to check in date. Users can enter wrong data in a hoard of ways. It is essential that all necessary actions must be taken to prevent entry of wrong data. So data validation requirements become important aspect of ancillary functionality.

6. **Response time functionality**—Sometimes, especially in real time applications, response times form part of core functionality. In business applications however, they form part of ancillary functionality. In web based applications, response times are important as the user and the server may not be at one location and if the application does not respond fast enough, the user may abort the application or do something else. Normally these form part of organizational development standards.

7. **Memory constraints functionality**—With software controlling every device today, the constraint still remains. It is a thing of the past as far as computers are concerned but devices like mobile phones, cars, washing machines and the like, not to mention rockets and space shuttles, memory constraints remains. In these cases, memory constraint has to be taken into consideration. These requirements can be obtained from hardware manufacturers who supply the hardware on which the software needs to function.

8. **Software footprint constraint**—When the software resides on a chip in small handheld devices and in various machines, the final size of the package that gets installed becomes very important. Now all the Computer Numerically

Controlled (CNC) machines not only in workshops but also in homes do have this limitation. In the present era, the cars, the refrigerators, the washing machines, the ovens, the mobile phones all have software on a chip which have limited capacity. This calls for restrictions on the size of the software that can be installed on the chip. This information can be obtained from the hardware manufacturers supplying the selected chips on which the software is going to be installed.

9. **Fault tolerance functionality**—Users make mistakes in using software, mostly unintentionally. This functionality ensures that software does not crash or abort when a mistake is committed by the user. It provides an error message and provides an alternative path and allows the user to use other functionality besides coming out of the fault-scenario smoothly without causing any damage.

10. **Reliability functionality**—There is a misconception in some quarters that since there are no moving parts, software ought to function reliably if it correctly runs once. But in these days of fast obsolescence, this becomes applicable. Every 3 years a new version of the operating system, browser, and middleware are released. The hardware, including processors, Network Interface Units (NIUs), switches, etc. are also upgraded every now and then. Since most modern applications work on the Internet the threat of viruses, spyware and malware also increased significantly. Any of these can impact the way our software works. Therefore, it is important to build the software in such a way that some of the possible changes do not affect the software or provide functionality to indicate that the environment changed so that corrective action could be taken. The users should not be faced with a software failure to understand that it needs upgrade. Software failure due to environmental change needs to be trapped and an alternative path needs to be available for the users for smooth changeover or closure of the application.

11. **Feel-good functionality**—This functionality is making the user interface look jazzy and sexy. The idea is to make the users feel good while using the software. This is adding glitz to the user interface screens and using nice pictures for buttons and icons and so on. I know of software product failure because the screens are not sexy!

12. **Esteem functionality**—This functionality brings pride to the users. For example, a Rolex watch delivers the same core functionality (showing time) as any other watch. But by ensuring that the thicker gold plating of the casing, scratch-proof crystal and so on, Rolex watches bring pride to the owners albeit at a higher cost. In software too this kind of functionality can be brought in the form of better messages, handling of error conditions, adding functionality that no other software possesses and so on.

13. **One-upmanship (competitive edge) functionality**—This functionality gives competitive edge to the software. It is having more functionality than any of the competitive software packages possesses. This functionality may be in core functionality or ancillary functionality. Compared feature to feature, this package would have one or more functional aspects in every feature over the competitive software packages.

2.3 Classification of Requirements Based on Product Construction Considerations

Any product, either for use by one organization or multiple organizations, needs to be built such that it works reliably without defects over the life of the product. In the case of software product, the life of the product is not dependent on its structure or wear and tear caused by functioning continuously but is dependent on the hardware platform it is installed on. A software product works as long as its hardware platform is unchanged, in working condition and is maintained well. In these days of fast obsolescence of hardware, the life of computer hardware itself is short compared to other types of machinery. As most of the present day software products are working on the Internet, there are many external factors that can adversely affect the application's reliability and defect-free functioning of the software product. Therefore, utmost care needs to be exercised in constructing the software product so that minor modifications in the environment outside the software and hardware platforms would not impact the reliability and defect-free functioning of the product. These requirements are unlikely to come from the end users or managements of the client organization. These need to be derived and implemented in the software product by the project team and the management of the software development organization. From this standpoint, the classification of requirements is enumerated below. Figure 2.3 depicts these requirements pictorially

1. **Maintainability**—The resultant product ought to be maintainable. That is, it ought to be possible to modify the code, add code or delete some code. The key aspect of maintainability is not just to add, modify or delete but to do those activities efficiently, effectively and with minimum expenditure of resources. Another important aspect of maintainability is that persons other than the original developers should be able to maintain the product. The product specifications that dictate product maintainability come under this category. These are normally in the standards and guidelines selected for the product.

2. **Flexibility**—Flexibility refers to the ability of the product to be used in multiple similar scenarios. For example, a materials management software product should be useful in engineering industry, chemical industry, mass production system and batch production system. Another understanding of flexibility refers to the ability of the product to be useful without any modification, when some of the underlying parameters change. For example, in a payroll software product, addition of a new deduction or addition of a new payment, should not render the product useless. On the other hand, the package should still be usable without changing the source code. The functionality specification that is focused on achieving flexibility in product functioning comes under this category. These specifications are usually part of standards, guidelines and ancillary functionality requirements.

3. **Efficiency**—The resultant product ought to use resources efficiently. The resources used by the product are not just computer resources (CPU, RAM, disk

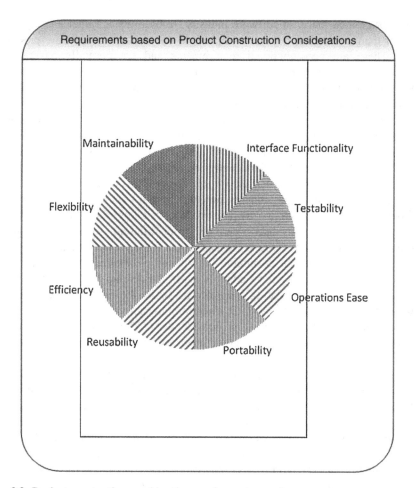

Fig. 2.3 Product construction consideration requirements

space etc.) but also the time of end users, bandwidth on the network, backup storage and so on. The product should minimize the time expected to be spent by the users, perhaps, by reducing the needed key strokes and mouse clicks. The functionality specification that is focused on efficiency of resource usage comes under this category. Standards, guidelines, ancillary functionality requirements provide these requirements.

4. **Reusability**—The product ought to be built is such a way that its components can be used in other products. Automobile industry implements this concept diligently. The same engine is used in multiple models. Components like a steering wheel, brakes, tires, shafts etc. are used without any change in many models. The dictum "There is no point in re-inventing the wheel" seems to

have originated there. In software industry, re-usability is rather an exception rather than a rule. But it pays to implement this concept in software products too. These are non-functional requirements and are normally covered by standards and guidelines or the organization.

5. **Portability**—In earlier days, porting is referred to as shifting the software developed in the same language such as COBOL from one hardware platform to another. But now, such a thing is passé. But another type of porting has come on to the scene. It is shifting the web site from one host to another. With cloud computing, it would be much more frequent in the future to be shifting applications from one host/data center to another. The functionality specifications focused on ensuring that the impact of porting is minimized come under this category and include standards and guidelines for achieving the portability.

6. **Operations ease**—Modern software products are large and multi-functional systems involving many users, perhaps, from different geographical regions spanning international borders. To keep such systems operational round the clock, they need specialists running the operations. Therefore, many of them need dedicated/shared systems administrators, DBAs and network administrators. In many cases, updating the software or hardware has to be achieved without bringing down the systems. Therefore, the software product needs to be built keeping all these aspects in consideration. These aspects are covered normally by the standards and guidelines dealing with product architecture, design and construction.

7. **Testability**—Of course, the product has to be testable and will be so. What is so special about testability then? It is generally agreed that 100 % testing of large software products is not practicable. Therefore, a variety of quality assurance activities are implemented during software development. The cost of fixing defects varies proportionately with the stage in which the defect is uncovered. A defect uncovered during unit testing costs much less to fix than a defect that is uncovered during system testing stage. The final product is always testable but what is sometimes becomes difficult to test is, the software unit/component. The software product has to be designed and built in such a way that every software unit is independently testable in a stand-alone manner. Testability requirements are normally covered by the standards and guidelines dealing with software architecture, design and construction guidelines.

8. **Interface functionality**—In these days of web based Internet applications, interfacing becomes essential. The internet itself is built with multiple layers. There are many browsers, and different servers, ISPs and networking protocols that an Internet application has to interface with. Additionally, the applications need to be built in such a way that it would be possible to interface with applications that the organization may build later on. This kind of functionality would be covered under standards and guidelines dealing with software design and construction, normally.

2.4 Classification of Requirements Based on Source of Requirements

Yet another way to look at requirements is based on the source from where it is obtained. There are many sources from which we can garner requirements for the proposed software product. Enumerated below are the possible sources for establishing the requirements for a software product.

1. **End users**—These people are those that use the end product to perform their individual business processes. The software product is basically aimed at fulfilling their needs. These people provide the core functionality especially relating to the aspects of inputs, process and outputs at working level. End users may not be able to provide the management requirements expected from the software. End users can be located in the case of project scenario (intended for use within one organization) in the departments funding the software development. In the case of a COTS (Commercial Off The Shelf) product scenario, end users are scattered across the target market for the product. We may perhaps need to conduct market surveys to get their needs or select randomly some end users and interview them to obtain their needs and the core functionality for the product.

2. **Management of customer organization**—These people provide the MIS (Management Information System) portion of the core functionality. They provide what information they need to extract from the software so that they can manage the organization effectively. These may include the special analyses, special reports, audit trails, security concerns, safety concerns and so on, necessary from the software product. In a project scenario, these people can be located in the organization looking at the organization chart. But in the case of COTS product development, we really need to put in efforts to locate such experts. They can be found in the domain experts, academia, and through market surveys selecting the senior management personnel to provide the information.

3. **Domain experts**—These individuals are those that have worked for long years in the business domain in which the proposed software product would be developed. These individuals are especially useful and extensively used in COTS product development. These people may or may not be IT (Information Technology) experts but they would have the knowledge of systems and procedures or the domain. They would know the detailed procedures, formats, templates, guidelines, standards and checklists used by the end users in the domain. Additionally they would be experts in the process that is used to convert the inputs to outputs as well as the legal issues involved with the domain. These people would be occasionally used in project scenarios to obtain information about industry best practices or when the requirements provided by the end users and their management are perceived to be either incomplete or ambiguous. Domain experts can provide end-to-end core functionality or clarify any issues thereof.

4. **Project team**—Project team comprises of the project mangers, project leaders, software designers, business analysts, programmers, testers, User Interface (UI) developers, and Database Administrators (DBAs). These people are also a source of providing requirements albeit the fact that they may not be able to provide core functionality unless the product is proposed to be used in software development activities. They would however be able to provide ancillary functionality such as usability, maintainability, safety, security, and reliability and so on.

5. **Statutes**—Statutes include governmental regulations pertaining not only to usability of the software but also about possible illegal activities. The intended software product shall not either commit, aid or abet any sort of criminal activity. Therefore, the requirements need to include all statutes that need to be implemented as well as ensure that all prohibited functionality is excluded from the proposed software product. The business analysts carrying out requirements analysis ought to be aware of the statutes that mandate inclusion of functionality as well as the functionality that is prohibited. Some of the examples of prohibited activities include stealing of personal data, siphoning away of monies in dormant bank accounts, sending spam emails and so on.

6. **Industry standards**—These include standards of industry (such as ISO, CMMI, NASSCOM (National Software and Services Companies of India) or professional associations (such as IEEE, SPINs, SPMNs) or organizational standards of either the vendor organization or the client organization. These standards address various aspects of software engineering methodologies including processes, guidelines, formats, templates and checklists). A host of information and best practices are available from such standards and ancillary functionality can be derived from those standards.

7. **Software designers**—Software designers can provide ancillary requirements about the efficiency, fault tolerance, operations ease, installation ease, usability, structural stability and so on of the end product. Software designers are also normally part of the project team but are treated separately because software designers play a key role in the final product. Finally it is the software designers that have to shoulder the responsibility for any missing/defective functionality in the end product of software development.

8. **Software programmers**—Software programmers are at the end of the chain for implementing the requirements in the software. However, they are the people who need to implement all the requirements conveyed to them using software design documents. However, coding guidelines, UI guidelines, and any other organizational guidelines would not form part of design documents and programmers are expected to be knowledgeable of such standards as well as implement them effectively in every program they code. Software programmers can specify ancillary functionality aspects pertaining to maintainability, testability, reusability of the code and so on.

9. **Software quality assurance team**—A Software quality assurance team includes reviewers, testers, and process specialists. These people can provide ancillary requirements about testability, and quality perspectives of the proposed software product.

10. **Management of software development team**—These individuals include project manager, project leader and other senior management personnel including program managers. These individuals would be in a position to have a bird's eye view of the overall project and would be able to provide interface requirements to ensure that the software product would be able to interface effectively with other applications in the organization. They would also be able to ensure that all functionality is included for the proposed software product.
11. **Marketing department**—Especially in product development organizations, marketing is a source of product requirements/specifications. A Marketing department can generate product requirements from the field staff or a market survey of the potential users of the proposed product. Market surveys are a very popular vehicle to collect user requirements and freeze product specifications. A Marketing department is the primary source for one-upmanship functionality in a COTS product.

2.5 Levels of Requirements

Institute of Electrical and Electronics Engineers (IEEE) has specified two levels of requirements specifications, namely, the User Requirements Specification and Software Requirements Specification. This gives rise to the question "Do we have two levels of requirements in software development?"

In any product there are two sets of requirements. One is the set of needs that the product fulfills and the second is the set of specifications that the product must adhere to in order for it to fulfill the needs of users. Let us take a bridge over a river (or a body of water) as an example. The need to be fulfilled is a bridge that can carry six lanes of traffic over a river of two furlongs breadth. Now based on this specification, some preliminary work is carried out to determine the depth of the water body, the banking on both the sides, the soil quality at the bottom of the water, the number of cars, trucks and other vehicles that traverse the bridge etc. Based on the study, the specifications for the bridge are finalized. These are the product specifications for the bridge, which include the load it has to support, the type of bridge (suspension, column supported, arched etc.), the width and so on. Based on these product specifications, a bridge is designed. Using the design, the bridge is constructed.

Similarly, in software development, the first expression of the need is the requirement of software to fulfill the requirements of a business process. Taking this first into account, a preliminary study is conducted to ascertain the needs of the business processes. From these results the product specifications would be drawn up.

In the manufacturing industry, the aspect of drawing up product specifications is referred to as Conceptual Design (or the High Level Design) and the working out the details is referred to as the Detail Design (or the Low Level Design).

Thus, there are three layers before the actual fabrication/manufacture/construction begins and these are:

1. The needs
2. The product Specifications
3. The design

In software development too, we have three levels before the coding begins. There are various nomenclatures for these three levels. I am presenting a few here but there could be others.

1. User Requirements Specification (URS), System Requirements Specification (SyRS), Business Function Specification (BFS), Functional Specification (FS), Requirements
2. Low Level Design (LLD), Software Requirements Specification (SRS), Functional Design Specification (FDS), Architecture
3. High Level Design (HLD), Software Design Description (SDD), Software Design Specification (SDS), Detail Design Specification (DDS)

Now the next question that arises is "what requirements do we need to manage—user requirements or software requirements?"

User requirements are original and first in the chain. Software requirements are not original. They are derived from user requirements. When user requirements change, the software requirements also change. Therefore, we need to manage user requirements. If we can minimize changes to user requirements, the changes to software requirements would automatically be minimized. This book focuses on managing user requirements.

2.6 Definition of Requirements in the Context of Software Development

How do you answer the question "what constitutes a software application?"

I am sure there will be numerous or multiple alternative answers for this question. The answer I select is that the application consists of a number of information processing processes. Information processing processes can be further divided into three classes, namely,

1. Input processes
2. Output processes
3. Associative processes

Input processes obtain information from outside the application boundary. The information would be provided by an entity (an individual, a machine or another computer application). The input can be data (facts, and figures about an entity), control data (triggers for events in the application such as start, stop, change, print etc.).

Output processes send information across the application boundary. The recipients of information would be an entity (an individual, a machine or another computer application). The output could be normal data in the form of a report either on paper or computer screen or to another application, or it could be control data in the form of trigger to another application. The giving/receiving application can be on another computer or another machine.

Associative processes are those processes that aid and assist the input and output processes in information processing. Consider a login process; it is neither an input nor an output. Similarly a file upload is a process; and so is an integrity checking process and so is the POD (Power On Diagnostics) process. These are all examples of associative processes.

Requirements elicitation/gathering is mainly enumerating all the processes that form a part of a software application and obtaining details about these processes so that downstream activities can be executed without further reference to the client or with minimal reference to the client.

Now each process has these attributes:

1. Inputs—each process receives certain inputs. While input processes receive inputs from external sources, output processes and associative processes receive inputs from the internal sources.
2. Outputs—Each process delivers some outputs. While output processes deliver to external recipients, input processes and associative processes deliver to internal recipients.
3. Process—each process carries out some transformation of inputs and converts them to outputs. Each process consists of some related steps with a start and an end event. The process includes verification of the inputs for their completeness, appropriateness and freedom from errors.
4. Triggers—each process needs a trigger that initiates the process into execution. The trigger could be an event initiated by an individual or another application or could even be by the application itself.

When we enumerate all the processes and define all the above four aspects for each of the processes comprising the software application, we have a complete requirements specification document.

2.7 Evolution of Requirements

Requirements start out as a single idea and over a period of time, evolve into a full set which can then be further processed into a software product. The phases in the evolution of requirements are discussed in this section. Please note that it is not mandatory that every organization uses these phases; some of the phases may be dropped or some other phases may be used; or they may use different set of phases altogether.

In a new product—non-existent in the market—This is a scenario in which a completely new product, the type which is not existing in the market is contemplated. In this scenario, the requirements are evolved as follows:

1. **Idea germination**—Here, the entrepreneur or the product manager germinates an idea based on his/her observations of the needs of the target market and perceives a need for a product that can fulfill the unfulfilled needs of target customers. In a large organization, there could be a few product managers and all of them could come up with new product ideas. It is stated that it takes about eight serious ideas to get one product idea to be approved and built. Not all approved ideas are successful and not all dropped ideas are bad ideas. Remember Chester Carlson was turned down for 5 years by organizations such as IBM and GE when he approached them for his idea on Xerox machines? This approved idea is the first phase in the evolution of requirements for developing new product the kind of which is non-existent in the market.

2. **Brainstorming**—Brainstorming is a technique in which experts in the subject at hand gather in an informal meeting and give a free rein to their imagination. All ideas expressed are recorded for later analysis which would shortlist ideas that are worthy of pursuing. In requirements management, all desirable product features are enumerated by the brainstorming, which are then sifted and feasible ones are culled. These form the initial requirements for the product.

3. **Market/customer/consultant surveys**—Now the initial requirements are tested using a market survey. Various methods of market survey are available and an appropriate one would be selected and used. The market survey would validate the initial requirements and usually adds a few more requirements. These requirements would be further validated by experts drawn from the market, consultants or academia using personal interviews.

4. **Personal interviews**—Personal interviews are conducted with selected experts who may be marketers, product designers, support staff, consumers, consultants or academics. The requirements finalized by the market surveys would be discussed with them for two purposes. One—to validate the requirements; two—to add to the requirements. This is the final step before attempting to design a prototype and go to market again to validate the product.

5. **Prototype and demos**—After the requirements are validated through personal interviews, normally a prototype of the product would be built. Prototypes are discussed in greater detail in Chap. 3. Now these prototypes are shown to prospective customers, and experts in the field. Their feedback is taken, evaluated and requirements are updated. This is the final step in the evolution of requirements.

6. **Freeze requirements**—Freezing the requirements involves documenting the requirements conforming to organizations standards and subjecting further changes to the rigor of configuration control and change management. The frozen requirements are then used to carry out full scale product design and development of the product and introducing it into the market. Normally, the

changes proposed on frozen requirements are considered for the next upgrade/
release of the product.

New product—existing in the market—This is a scenario in which the
product is new to the organization proposing to develop it but something similar is
available in the market. When somebody or an organization wishes to develop a
competitor to a product that is already existing in the market, the requirements
evolve through the following phases:

1. **Idea germination**—The entrepreneur or the product manager or someone with
   a say gets an idea to develop a product as a competitor for an existing product.
   The existing product may not be fulfilling the market expectations or the market
   is large enough to accommodate a new and innovative product or some such
   motive could be behind the idea. This is the preliminary requirement.
2. **Market/customer/consultant surveys**—Market surveys are conducted to
   confirm the need for an additional product and to unearth the needs unfulfilled
   by products existing in the market.
3. **Personal interviews**—Personal interviews with experts in the field are conducted to validate the data from market surveys and to add any more
   requirements to the list.
4. **Brainstorming**—In the brainstorming, one-upmanship ideas over the existing
   product are generated. Ideas about more functionality, better presentation,
   better workflow, improved ease-of-use, more user options, flexibility etc. are
   generated during brainstorming. These would be analyzed and requirements are
   finalized.
5. **Prototypes**—Normally prototypes are not constructed in this scenario as a
   working product is available in the market. But, it may be used sometimes to
   prove a concept or a feature and get feedback from the market.
6. **Freeze requirements**—Freezing requirements as noted earlier involves
   approving the requirements document and subjecting it to the rigor of configuration control and change management.

Product upgrade—We have a product that has been in the market for some
time and we have been receiving feedback about the desired additional features or
improvements in the existing features from our customers, field support staff,
marketers and Value Added Resellers (VARs). We also find that competitors have
brought their products into the market which are cutting into our market share.
Therefore, we wish to upgrade our product to keep it competitive and attractive to
the market. Here is how the requirements evolve:

1. **Feedback/Surveys from VARS**—VARS are one valuable source of feedback
   about the existing product owing to their proximity to the competitors in the
   market and end users. Whenever they provide feedback, we need to analyze and
   resolve it. When we contemplate upgrading the product, we need to conduct a
   survey to elicit their views on the improvements desirable to our product over
   and above what they already communicated. The information obtained from the

feedback and survey of VARS needs to be analyzed to remove duplication and the feasibility of implementation.

2. **Feedback/Surveys from tech support staff**—Support staff is in the field and are the first contact with the customers/end users. They would receive issues, concerns and suggestions for improvement from the customers. Most product organizations would have a formal mechanism to capture this kind of information during their interaction with customers. When we contemplate a product upgrade, we need to collate all such feedback and conduct a formal survey to elicit any further suggestions from the field staff to comprehensively capture all the expertise gained by the field support executives. The feedback and survey results can be analyzed to finalize upgrade requirements.

3. **Customer/market surveys**—Customers are the only people in the supply chain that can provide first-hand feedback. All others can only provide second-hand feedback. Therefore, we need to conduct customer surveys. These would validate the feedback obtained from VARs and field support staff.

4. **Personal interviews**—We need to conduct personal interviews to validate the findings of various surveys conducted as well as to uncover any biases or prejudices that have crept into our survey results. We conduct personal interviews on a sampling basis using stratified sampling technique.

5. **Freezing of requirements**—Once we have collected feedback/survey results from VARS, field support staff and customers, we analyze the results to consolidate the requirements and eliminate duplicates. We select the requirements for the product upgrade based on their feasibility. Then we document the requirements conforming to organizational standards and subjecting the document to the rigor of configuration and change management.

Project development scenario—The end result of software development either in the product development or project development is a software product. Then why should we distinguish product development and project development? The way I distinguish project development from product development is based on the use of the end product. In the *project development* scenario, the end product is proposed to be used by one customer or one set of end users within one organization. The end product of *product development* on the other hand, is proposed to be used in multiple locations, in multiple organizations, and by different sets of end users. Because of this distinction, the process of obtaining and finalizing requirements assumes different levels of importance. In a product development scenario, we need to spend much more time on finalizing requirements because any mistake committed during this phase would have a strategic impact on the final success of the product and the survival of the organization itself. Since the project is for one customer, the preferences of that customer assume paramount importance. In the project development scenario, we do not have VARS or field support staff or multiple customers to cater to. We have only one customer and one set of end users to whom we should listen to and satisfy them through our software product. In this scenario, we have two classes, namely in-house project and outsourced project. In the in-house project, the end product would be used within the same organization, even if it is in another department. An outsourced project is one in which the end product would be used in an organization other than the one in which it was developed. In outsourced projects, the organization that develops the product would normally be specializing in software development for other organizations. In both scenarios, the business analysts would interact with the end users performing business processes, to elicit, gather and finalize requirements. The evolution of requirements in these two scenarios is discussed below.

In-house project—new project—The evolution of requirements in in-house new project would take this course:

1. **Proposal by a functional department**—A department responsible for a set of business processes proposes computerization of the processes with a view, perhaps to increase efficiency, reduce turnaround time or improve quality and initiates the process. The first step in the process is the definition of the scope of work and defining the boundaries of the application that is proposed for computerization. This is the initial requirement. This scope would then be forwarded to the IS (Information Systems) department to come up with a budget. The department obtains financial approval for the budget requested by the IS department and requests the IS department to carry out all necessary activities to computerize the selected set of business processes. The IS department may be called by different names in different organizations.
2. **User Requirements definition**—Business Analysts from the IS department would elicit/gather requirements from the end users. Elicitation and gathering of requirements is described in Chap. 3 of this book. The principal methods used in this scenario are personal interviews and study of procedure manuals, forms and templates used in the performance of the processes as well as personal observation to collate the user requirements. All the requirements collated are analyzed to finalize requirements for the proposed system. Requirements Analysis is described in Chap. 4 of this book.
3. **Finalization of the requirements**—The requirements allocated to the proposed system are then documented conforming to the organizational standards. This document is internally reviewed and approved within the IS department and is then forwarded to the functional department for review and approval. The functional department reviews the document and requests clarifications, if necessary. Normally a meeting takes place between the executives of the functional department and the IS department, in which the document contents are discussed and clarifications from both sides are provided. The IS department implements any feedback received from the functional department and obtains the approval from the functional department. This document acts as the reference between the IS department and the functional department during the course of the software development. This document would be subjected to the rigor of the configuration and change management.

In-house project—upgrade—Usually, the software products being developed and used inside an organization are maintained regularly as needed. The strict meaning of "maintenance" is to restore/keep an artifact/product to its working state. However, in the software industry, product expansion is also carried out under the caption of "maintenance". But such product expansions, carried out as part of software maintenance, are usually minor in nature. When the product is scaled up significantly, an upgrade project is undertaken. Such major upgrades are necessitated from time to time and could be to use the newer technologies (such as web enabling the applications) or newer business scenarios (such as mergers or acquisitions or a major re-organization, etc.) For an upgrade project undertaken in the organization, in-house, the requirements would evolve in the following manner.

1. **Proposal by the functional department**—The trigger for the functional department to propose a major upgrade to the software product would be external usually. The trigger may be availability of new technology, availability of sparable funds, a change in the business scenario and so on. In such cases, the functional department would define the scope of work for the product upgrade and invite the IS department to propose a budget and obtains the financial approval for the budget. The definition of the scope of work for the expansion would be the initial set of requirements.

2. **Use Requirements definition**—Business Analysts would approach the executives of the functional department and elicit/gather the requirements. In this case, the elicitation/gathering is limited the expanded functionality only. If the upgrade is to convert the existing application from an existing platform to a newer platform, this phase may be avoided altogether except to see if any additional requirements are necessitated due to the change in the technical platform. These requirements are analyzed and requirements are finalized for the project.

3. **Finalization of the requirements**—These requirements are documented and are forwarded to the functional department for approval. The functional department accords approval after getting its feedback implemented in the document by the IS department. This document is then subjected to the rigor of the organizational configuration and change management.

Outsourced project—new project—The outsourced project in this context is the project executed at the vendor's place for a client. An organization realized the need for computerization of a set of its business processes, but decided to outsource the software development portion of the project to a specialist software development organization. Now, this project is proposed to be executed at the organization specializing in software development. The requirements evolve in the following manner in this scenario.

1. **Project acquisition**—The first step in the project execution is to acquire the project from an outsourcer organization. The projects are outsourced in different combination of software development phases, such as (a) Requirements, design, construction, testing and delivery; (b) Design, construction, testing and delivery; (c) Construction, testing and delivery; (d) Testing and delivery and any other combination. The projects may be on fixed-price contracts or time-

and-material priced contracts. Whatever the case may be, the project acquisition phase would certainly include the definition of the scope of the project, which is the initial requirement for the project.

2. **Requirements elicitation/gathering**—During this phase, Business Analysts from the vendor organization (or internal analysts if requirements definition is not outsourced) would elicit/gather the requirements from the client executives. They would be using personal interviews and surveys to elicit the requirements and study the existing process manuals, formats and templates to gather the requirements. They would then be analyzed to form the project requirements. Sometimes the requirements may be provided by the client along with the purchase order. In such cases, the organization just needs to assess the received requirements for adequacy and completeness.

3. **Analyze the requirements**—Subject the collated requirements to analysis to determine their technical feasibility, eliminate duplicates, group them into logical groups and prioritize them, Analysis is described in Chap. 4 of this book.

4. **Finalize the requirements**—This involves documenting the analyzed requirements to ensure that they are conforming to the agreed standards and obtaining internal approvals after due quality assurance process of the organization.

5. **Customer approvals**—The finalized requirements would then be submitted for customer review and approval. The customer would review the requirements document and request improvement, if necessary, by providing the feedback. When the customer is satisfied, approval would be accorded to the requirements document. Now, this approved requirements document would be brought under the vendor's configuration and change management process. This document would be the reference point for all interaction between the customer and the vendor in matters relating to project requirements.

Outsourced project—upgrade—Outsourcing a major product upgrade to a vendor is risky as the source code needs to be provided to the vendor to upgrade the product. It had to be done in the case of Y2 K conversion. Millions of LOC (Lines of Code) had to be uploaded to vendor's computers for them to convert it to be compliant with Y2 K requirements. No major breach-of-trust case was reported arising out of stealing intellectual property contained in the source code even though most of the upgrade is carried out overseas. In cases where the upgrade is to re-develop the code on a new platform, of course, the original source code need not be given to the vendor. Requirements in this scenario would evolve in this manner:

1. **Project acquisition**—In this scenario, the definition of scope would be much more lucid than in the scope definition of a new project. This is due to the fact that the product is working and the upgrade requirements are better defined. It is usual to get a list of detailed requirements in the RFP (Request for Proposal) itself. However, in platform upgrade, requirements may have to be elicited/gathered from the customer executives. The RFP provides the initial requirements.

2. **Requirements elicitation/gathering**—When required, the requirements need to be gathered in the same way as explained for the new project requirements evolution in the above section. When requirements need to be elicited/gathered, they will be collated using personal interviews and surveys as the tools.
3. **Requirements analysis**—When requirements are provided, they will be analyzed for their adequacy and completeness and clarifications are obtained. If there are requirements which are elicited/gathered, they will be analyzed for their feasibility for implementation.
4. **Finalize the requirements**—The finalized requirements would then be documented conforming to the agreed standards. These will be subjected to an internal quality assurance process and internal approvals need to be obtained. Then the approved document would be forwarded to the customer for their feedback and approval.
5. **Customer approvals**—A Customer would review the requirements document for the completeness of information and to ensure that the requirements are properly understood by the vendor. The customer would communicate feedback, if any, for implementation and after the document is to their satisfaction, the customer would accord the approval to the requirements document. This approved document would be subjected to the rigor of the vendor's configuration and change management process. This document would be the reference point for all matters relating to project requirements between the outsourcer and the vendor.

Of course, there would be variations to the evolution of requirements from one organization to another. What is presented here are the typical scenarios. Another aspect to be noted is that the projects are of various types. There are full life cycle projects, part life cycle projects, testing projects, conversion projects, porting projects, migration projects and so on.[1] Whatever the project type, requirements are the first step in the project execution and any error committed in this phase would have a recurring impact on all the subsequent phases of the project.

---

[1] Interested readers are suggested to read the book "Mastering Software Project Management: Best Practices, Tools and Techniques" by Murali Chemuturi and Thomas M. Cagley, Jr., published by J.Ross Publishing, Inc, USA, 2010.

Chapter 3
Elicitation and Gathering of Requirements

3.1 Introduction

The Dictionary meaning of the term "elicit" is to "draw forth or bring out" something that is latent or potential or "call forth or draw out" as information or response. This connotes a dialog in which information is drawn out from a party possessing the needed information.

CMMI v 1.3 defines elicitation as *"Using systematic techniques such as prototypes and structured surveys to proactively identify and document customer and end-user needs"*. This definition too indicates a dialog between software developers and customers albeit using of techniques like prototyping and surveys to draw out the needed information.

The Dictionary assigns multiple meanings to the term "gather". One of them is "to bring together" as in "tried to gather a crowd". Another meaning is to pick up or amass as if "by harvesting/gathering ideas for the project". Another one is "to effect collection of" as in "gather contributions". As you can see, the term "gather" connotes collecting things which are available but scattered over the place.

While most technical documents on requirements management combine both elicitation and gathering together, they are distinct from each other.

Elicitation is first hand collection of information from individuals who are directly concerned with the project, using interviews. It is from primary sources. Primary sources include end users, experts, and brain storming.

Gathering is an indirect collection of information from sources other than human beings. It is from secondary sources. Secondary sources include documents, existing applications, and standards and guidelines.

Elicitation and gathering of requirements is the precursor for all requirements management activities. Both these techniques are widely used by the software development industry. In some cases, elicitation precedes gathering and in some cases gathering precedes elicitation. In a few cases only one (either elicitation or gathering) technique may be used. Scenarios, in which neither of these techniques is used, may exist, but rarely. It may happen that the customer provides a comprehensive set of requirements to the development team in which case, the team can start software design activity. Normally, the development team may have to add some ancillary functionality to the requirements provided by the customer before embarking on the software design activity. It is possible that elicitation may not be used but some sort of gathering needs to be used in software development activity, especially when the customer provides a comprehensive set of core functionality requirements.

We collect the following information for each of the processes, to be able to analyze it and prepare a requirements specification document that can guide the downstream activities of software design, construction and testing.

1. The list of processes within the application boundary
2. The four attributes for each process, namely, the inputs, the outputs, the transformation and the verifications carried out on the inputs
3. The trigger for the process
4. The exit point for the process

Now let us look into each of these techniques in a detailed manner.

3.2 Elicitation of Requirements

As noted earlier, elicitation is obtaining information from primary sources of information. Merriam Webster's Dictionary (http://www.merriam-webster.com) provides the following definitions of the word "elicit" among others:

1. "To draw forth or bring out (something latent or potential)"
2. "To call forth or draw out (as information or response)"

The examples, among others, provided are:

1. She has been unable to *elicit* much sympathy from the public
2. My question *elicited* no response

The connotation that is implied in the term "elicitation" is that information is obtained using personal interviews as the vehicle for obtaining the required information.

The following techniques are used to elicit requirements.

1. Personal Interviews
2. Questionnaires
3. Customer/market surveys
4. Observation
5. Demonstration of product prototypes or the product itself
6. Brainstorming

Now let us look at each of these techniques in a little more detail.

3.2.1 Personal Interviews

Personal interviews are perhaps most extensively used especially in the case of project software development.

When software is developed in-house, the Business Analysts from the IS department meet the end users to collect requirements using personal interviews. Personal interviews are the primary means of capturing project requirements in the case of in-house software development projects.

When software development is outsourced, the vendor receives a set of top level rudimentary requirements which need to be developed further so that software design can begin. The development team studies the preliminary requirements and identifies gaps–gaps between what is provided in the preliminary requirements and the granularity needed for commencing the software design. It is sometimes possible to bridge the gaps using other methods like emails and phone calls when the identified gaps are minimal or are of simple nature. But when the gaps are significantly many, or are of complex nature (that is, the development team is not able to understand fully the implications of the preliminary requirements received), personal interviews become necessary. Unless proper care is taken, it is easy to waste time and effort on eliciting information using personal interviews. The following steps would aid in the collection of right information using personal interviews:

1. **Planning of personal interviews**—Planning involves identifying the resources required to achieve the objective. In this case, it is the concerned executives who can provide necessary information, time and effort required to collect the information as well as logistics necessary to conduct the interviews and to sift the information collected. We begin by making a list of individuals that can provide relevant and useful information. We then estimate the clock hours needed to elicit information from each of those executives. We also estimate the time and effort required from us to prepare for conducting the interview and the type of persons that can conduct the interview and elicit the requirements.

2. **Study the preliminary information and relevant subject literature to understand the domain at hand**—If we are already knowledgeable in the domain at hand, we may perhaps skip this step. Otherwise, it pays to study the preliminary information provided by the client and identify the gaps. If there is no preliminary information from the client, we may study the literature available on the subject and this study could include running demos of similar products developed earlier in our organization or demos available on the Internet. This will help us in becoming familiar with the jargon of the interviewee and aid in quick understanding of the responses received in the interview. This is a very important step in eliciting requirements using personal interview technique. Failures in eliciting requirements using personal interviews mainly stem from skipping this step or performing this step in a very cursory manner. A well prepared interviewer can obtain all the information in

one iteration where as an ill-prepared person would need multiple iterations to obtain the same amount of information.

3. **Prepare a set of questions to ask to aid in the structured elicitation of information**—Whether it is from our existing knowledge of the domain or from the study of literature and demos of similar software products, as well as from the gaps identified, we need to prepare a set of questions to guide us in conducting each of the interviews effectively. We need to prepare questions in such a way that each question would elicit information about one topic at a time. We can also ask corollaries as necessary to ensure completeness of the provided information. We record the questions in a format or template designed for conducting the interviews so that we can log the information received against the relevant question. It pays rich dividends if we get the question set to a peer review and implement the feedback and thereby improve the quality of questions. On each topic, in order to have complete information, we need to ask:

   a. The trigger that activates the process
   b. The inputs for the process (data items)
   c. The outputs/deliverable of the process (data items)
   d. The validations that need to be carried out on the inputs to ensure that proper inputs are fed to the process

4. **Prepare formats and templates to capture information efficiently**—We need to prepare the formats and templates for capturing the information effectively during the course of the interview. It is tempting to go into a personal interview with only white paper scribbling pad, pencil and an eraser. Doing so, we would waste, not only our time but also of the interviewee. It is normal in these days to use a laptop for capturing information during interview. Paper-based templates and formats are passé. Create special folders for capturing the information. Copy the formats and templates into those folders. Fill in the prepared questions into those formats and templates as appropriate. Arrange a peer review to ensure preparedness before we actually begin the interview. A suggested template for capturing the process information is given as Table 3.1. This may be used while carrying out personal interviews to capture the process information.

5. **Fix appointments with the identified executives**—most executives providing information for capturing the requirements do prefer to plan the session so that they can arrange their schedule and ensure that no interruption is caused. You may call the executive and fix up the date and time of interview. Sometimes, you may need to speak with the secretary, especially when you are trying to interview senior management personnel. You may use email or a corporate calendar also to fix up an interview, if the concerned executive is comfortable with such an arrangement. In today's automated environments, you may need to put up a request for an appointment on the organization's calendar system. Whatever be the method may be, ensure to set-up the appointment so that when you arrive for the interview, the executive is ready to give you the information.

Table 3.1 Template for capturing process information

Interview Template for eliciting information through personal interview	
Name of Business Analyst	
Function Id	
Workstation Id	
Mode of information capture (Personal Interview/documentation study/user supplied/ any other)	
Date of information capture	

Note: One function (process may contain multiple inputs, multiple outputs but shall contain only one process. You may use multiple sheets to capture inputs and outputs, but process will have only one sheets.

Inputs (use a separate sheet for each input)

Input Id	
Input Description	
Agency from which it is received	
Trigger for the input (describe the event that triggers the input or periodicity of receipt)	
Contents of the input	
Description of verification to ensure completeness and appropriateness of the input	
Exit point – when the input process can be termed successfully complete	
Abort points – when the input process can be terminated without completing it	

Outputs (use a separate sheet for each output)

Output Id	
Output Description	
Medium of output (paper/screen/on wire)	
Agency to which it needs to be sent :	
Trigger for the output (describe the event that triggers the generation of output or periodicity of generation)	
Contents of the output	
Description of verification to ensure completeness and appropriateness of the output	
Exit point – when the output process can be termed successfully complete	
Abort points – when the output process can be terminated without completing it	

Table 3.1 continued

Process	
Process Id	
Brief description of the process	
Objectives to be achieved by the process	
Process performers	
Trigger for the process (describe what event/time initiates the process)	
Process steps (attach a flow chart/data flow diagram)	
Exit point (describe how the process can be terminated after successful completion)	
Abort events (describe the events when the process can be terminated midway)	

Data description – enumerate the data that is used by the process

Name of the form :
(Use additional tables for each form)

Id of data item	Data type (numeric/character/ control)	Maximum length in characters
1		
2		
3		
n		

Having an appointment also enables the executive to be prepared for the interview as well as to make the templates/formats used by the company available to you.

6. **Conduct the interview effectively**—It is often the case that the personal interview goes astray and takes much more time than envisaged. It is easy to get distracted from the main objective of the interview which is to obtain information about the processes being performed at the workstation. Since human beings are involved, empathy and gentleness are required while conducting the interview. Here is where preparation about the processes helps in gently questioning and obtaining the right information. When you conclude the interview, you must have all the information (inputs, outputs, process steps, and data items) of all processes performed by the person. Here are some suggestions on conducting an effective interview:

a. Be on time so that the executive would not be kept waiting.

b. Ask questions in the gentlest manner possible and listen attentively and carefully so that the need for repetition is minimized

c. Make notes as you listen and understand the explanation. If you are recording the interview, inform the interviewee of the fact as some people may object to being taped. The issue with audio recording is that we cannot quickly check if we have all the information at the end of the interview. To do so, you need to playback the entire interview! So, as you record the interview, be careful to capture all the required information.

d. Draw rough flow charts to clarify your understanding of the process steps and get them confirmed by the interviewee.

e. Never dispute or pick an argument with the person. If you get the feeling that the person is not right, you can get it clarified from his superior or with the same person in a second round. But if you get into an argument, the interview goes astray.

f. When in doubt, describe your understanding and ask the individual to confirm your understanding rather than ask the person to repeat.

g. In some cases, it becomes necessary to have another round of discussion, fix up the next appointment before you conclude.

h. Do not forget to collect the forms used by the individual. These help you to understand the person's inputs and outputs.

i. When you collect the forms or note down the data items, ask for the type of data and its maximum length. Non-software personnel are apt to give the average length, but it is necessary to provide for the maximum length in the database.

j. When you have to terminate an interview before capturing information about all the processes performed by the individual, terminate after capture of complete information about a process. Terminating midway for a process tends to repeat information capture for the process all over again.

7. **Capture information**—While conducting the interview, the information provided by the individual may be captured using the template provided in Table 3.1. Capturing information while the interview is in progress, may force you to hurry in noting down the details. Unless one is adept in speed writing, it would be difficult to capture the information provided by the interviewee by hand. It may be a good idea to use a tape recorder of some kind or use a laptop's sound recording facility to capture the voice of the interview so that it can be replayed later, to capture information into the template.

8. Check for completeness of information using organizational checklists

9. Sift the information collected into requirements, information for filling in the identified gaps and data useful for software design such as workflows.

3.2.2 Customer/Market Surveys

This is perhaps the most extensively used technique to capture the requirements for a product and is used widely in the software product development scenario. Surveys generally collect information from a wide audience. The objective of a survey can be to obtain information, or prove/disprove a hypothesis, to learn the social trend, to learn how a particular product is being used and so on. The survey used in requirements management is to obtain information from the targeted users of the proposed product. The information thus obtained shall be analyzed to develop the requirements for the proposed product. Surveys elicit opinions from both the existing customers and the people forming part of the target market. Normally the following categories of persons would be approached to provide the information:

1. CIOs (Chief Information Officers) or whoever is looking after the computer departments of the organization, in case the proposed product is focused on assisting the information departments in efficiently managing the IS (Information Systems) functioning.
2. The general public, if the proposed product is targeted at the public at large.
3. Existing customers, if an existing product is proposed to be upgraded with new functionality.
4. VARs (Value Added Resellers) both for developing a new product or upgrading an existing product. For a new product, the VARs would provide valuable information based on their experience with the other products in the market.
5. Internal product technical support staff for improvements to solve the problems they faced in the field.
6. Consultants of specialized fields to provide information about the product requirements in their specialty.

Normally surveys are conducted in three ways:

1. **Face-to-face method**—in this method, an individual representing the organization would approach the members of the selected audience and personally interview the person to obtain necessary information. This would be used, sparingly as it is very costly compared to the postal survey. When it is expected that the responses to a questionnaire would not be satisfactory and supplemental questions are needed to elicit complete information, this technique would be used. The flip side of this technique is its high cost as higher cost individuals are needed to interview and obtain the information. Another disadvantage is that the number of responses would be far less compared to postal surveys. It is still used for a few cases to check the efficacy of the findings of a postal survey. This technique uses the methodology described in Sect. 3.2.1.
2. **Postal method**—in this method, questionnaires are mailed/emailed to the target participants and responses are collected. When mailers are used, it is customary to provide addressed and stamped envelopes to mail back the responses so that the respondent is not required to spend money for mailing back the response. Questionnaires are detailed in Sect. 3.2.3.

3. **Web based surveys**—This is the latest method being used by software product developers. The main advantage of this method is that collation of information from the survey is automatic and immediate. In both the above methods, namely, face-to-face and postal surveys, the results need to be entered into a database to enable analysis. In this case, the results are automatically and immediately stored into the database and the turnaround time is drastically reduced. Some precautions are necessary to limit the participants to a geographic region. Otherwise, we may receive responses from all over the world. In the case of face-to-face and postal surveys, the participants are automatically limited to the selected set or respondents. But in web based surveys, we need to devise mechanisms to limit the participants. Often, web based surveys are coupled with email invitations to participate in the survey so as to limit the participants. In addition to the cost of developing the questionnaire, the cost of the software necessary to conduct the survey would be involved. Alternately, we may use specialized providers of web based surveys, who, fortunately, are available and provide the service at an affordable cost. This technique also uses questionnaires detailed in Sect. 3.2.3.

3.2.3 Questionnaires

Questionnaires are one of the very popular methods of eliciting requirements. These are used mostly in the software product development scenario. The product management team or whoever is looking after the development of requirements for the proposed product designs a suitable questionnaire. Normally, most of the product development scenarios are based on an existing product. Therefore, the questionnaires enumerate all the existing features of the product as well as new ones and the respondent is requested to:

1. Set the order of importance of the features
2. The features that are most essential
3. The features that are not essential
4. The features that may motivate the respondent to either buy or switch to the new product
5. Any other additional features the respondent would like to see in the new product.

Sometimes, an incentive to respond is provided to obtain maximum response for the questionnaire. Normally the incentive would be a draw of lots for a prize or a free report or something like that.

In a project development scenario, when requirements are elicited by personal interviews detailed in Sect. 3.2.1 above, questionnaires could be used to aid the interviewer to obtain full information from the interviewee. Here, the questionnaire is not administered as in an examination with the respondent filling in the appropriate answers. Here the questionnaire acts as a sort of tickler so that the interviewer extracts all the information available with the interviewee.

3.2.4 Observation

In some cases, the analyst simply observes the operations while they are taking place and collects the information. Take for example,

1. the point of sales in a super market/mall
2. a busy bank teller
3. a guest registration counter in a busy hospital/hotel
4. an online registration/reservation system,
5. a customer support person/help desk
6. an enquiry/assistance counter in a busy public facility including bus stations, railway stations, airports etc.
7. working machines in the case of real time software development
8. or any other such scenario.

In all the above scenarios, the observation method comes in handy.

The counter executive would be able to explain his/her version, but there is another side that of the customer/patient, who are not part of the organization. Customer/patient are external to the system but are impacted by it. If the system does not take their concerns into consideration, it may turn them off from the organization. Customers' views can be obtained in such cases, but observing a few would be helpful. Personal observation supplements/confirms the information obtained by surveys and personal interviews. Personal observation is also helpful in confirming the efficacy of the system after pilot implementation so that improvements, if any required, can be implemented before the final roll out. Observation helps in obtaining the following information, first hand:

1. Response times needed from the system
2. Ease of use of the system
3. Efficiency of the system in practical use
4. The additional help needed by the users while utilizing the system
5. Productivity of the end-user.

The information obtained from personal observations would be leading to ancillary functionality requirements rather than core functionality requirements.

3.2.5 Demonstration of Product Prototypes

Demonstration of product prototypes is more often used to finalize requirements than to obtain original requirements. Often times, the customer is not willing to sign off on a requirements document as they are not well versed with the software development methodologies and techniques such as DFDs (Data Flow Diagrams), ERDs (Entity Relationship Diagrams), Use Cases, Class Diagrams etc. They are afraid that they are approving something they do not understand and are unsure if their stated requirements are included. Therefore, software development organizations often build a prototype of the proposed system as they understood and demonstrate it to the end-users. The end-users look at the prototype and provide feedback on:

1. the missing features, if any, from the stated requirements
2. modification of the implementation of their requirements to suit their needs
3. wrong implementation of their requirements, if any
4. any additional requirements that were not included earlier

Prototypes used for this type of demonstrations are of two types,

1. **Use and discard prototypes**—these are built using a graphic tool like Visio or a presentation tool like PowerPoint, or a spreadsheet like Excel. Once all the requirements are obtained, the prototype would be discarded. That is, it would serve no other purpose than to give an idea to users and obtain their feedback and approval.
2. **Use and improve prototypes**—here, the prototype would be a mock up on the actual development platform. The skeleton of the product would be built with not much code behind the screens or reports but enough to demonstrate the product to the end users. It would consist of screen layouts and report formats. Screen layouts enable a user to visualize how the information input/enquiries would look like and to determine if they meet their requirements. Report layouts enable them to analyze the information outputs and to determine if their requirements are understood by the developers in the right perspective.

When the initial requirements are very fluid and unreliable, use and throw prototypes are used. But if, the initial requirements are more reliable, use and improve prototypes are used. However, it is the project team's choice to use either type of prototype. One additional advantage of prototyping is that the software design also gets finalized by the time requirements are finalized. The concomitant disadvantage is that effort must be expended on software design right at the time of eliciting requirements.

Prototypes are used in project development scenarios largely. Prototypes are also used when developing a new product the like of which does not exist in the market.

3.2.6 Product Demonstrations

Now a day, quite a few COTS products are available with comprehensive functionality with built in best practices for such areas as ERP (Enterprise Resources

Planning), CRM (Customer Relationship Management), SCM (Supply Chain Management) etc. A number of organizations are using these readymade products to automate their operations. But often times, some amount of customization in the product would be needed to make it suitable to the unique needs of the specific organizations. In these cases, demonstration of the product would be used to elicit requirements for the customization. In this case, a Business Analyst proficient in the product as well as the functional domain would guide the audience through all the features of the product. The users would then point out the areas that need customization to suit their organization. These aspects would be noted by the Business Analyst for later analysis and freezing of the project requirements.

3.2.7 Brainstorming

Brainstorming has a variety of uses but in the context of requirements engineering and management, it is used mainly in generating initial requirements in the product development scenario. For a new product that is not existing in the market, it would aid in enumerating the features and for an existing product, it aids in enumerating the additional features. In brainstorming, a group of people knowledgeable on the subject gather in an informal environment and air their views on the subject at hand. The idea is to give wings to each person's imagination so that the best information can be brought out. No one in the group would criticize other's view except to the extent of pointing out any blatant mistakes. All views are recorded for later analysis. The analysis would sift out the grain from the chaff and cull the right requirements for use.

All the requirements captured through elicitation are analyzed. Analysis of requirements is covered in Chap. 4. The analysis would shortlist the requirements for the project. These requirements combined with other requirements from other sources including gathering of requirements would be documented and finalized for the project.

3.3 Gathering Requirements

Merriam Webster's Dictionary (http://www.merriam-webster.com) defines the word "gather" to mean "to pick up or amass as if by harvesting (gathering ideas for the project)"; "to scoop up or take up from a resting place" among others for using that word as a transitive verb. As examples, it gives:

1. "Give me just a minute to *gather* my things and then we can leave"
2. "The child is *gathering* flowers to give to his mother"
3. "The police are continuing to *gather* evidence relating to the crime"

I gave these examples to enunciate the difference between "elicitation" and "gathering" of requirements in the context of software development. Elicitation is to obtain by enquiry. Elicitation is the first hand collection of the information from either the individuals performing the business processes or by observation. On the other hand, gathering connotes that something is already available, scattered or lying around and that it is there to be collected. Thus, gathering, basically, is collecting information from secondary sources which are published materials, records, process documents and so on. In requirements management, both elicitation and gathering techniques are used to collate information to analyze and finalize requirements for the proposed software product. Even in organizations that do not use documented processes to guide the performance of business processes, Business Analysts need to refer to formats and templates for gathering information about the formats of the reports and enquiries to be included in the proposed software product.

The following are some of the documents that we study to gather project requirements :

1. **Organizational records**—These are logs of activities carried out in the organizations. They give information for the design of database, maximum lengths to be provided for various fields, the size to which the data can grow over a period of time and so on. The following information becomes available from organizational records:

   a. Number of tables that may have to be designed—each type of record can roughly correspond with a table in a database. Sometimes, for example a purchase order in a material management application, can spawn out to more than one table.
   b. Number and type of fields in various tables—each column in a record can roughly correspond with a filled in table. It is also common in manually maintained records to include more than one data item in one column (such as address) against a row, in which case, each distinct data item corresponds to a field.
   c. Maximum lengths for each of the fields—Often times, the end users specify the average length for a field than the maximum that may occur. Records are one good source to determine the maximum length necessary to hold data in a field.
   d. Relations between various tables—study of records would reveal the information drawn from other records which could help in determining the relations between tables in the database.
   e. Size of tables—the number of records gives an idea of the number of records and the time frame in which they accumulated. Extrapolating these numbers, we can determine the size to which each table can grow over a period of time. This information coupled with the organization's data retention policies can help us determine the database growth strategies.

2. **Process documentation**—Organizational process documentation would usually consist of processes, procedures, standards, guidelines, SOPs (Standard Operating Procedures) with processes at the highest level in each category. Studying these documents would give us a complete idea about the organizational functioning and how the business processes would be performed by individuals designated for each process. When an organization has a well documented process which is followed scrupulously in the organization, the need to conduct personal interviews would be minimized. They provide almost all the information necessary to finalize the requirements.

3. **Standards, and guidelines**—Standards and guidelines certainly form part of process documentation but deserve a special mention because they mandate how business processes are to be performed in the organization without exceptions. Standards, and guidelines give us the idea about the business processes of the organization and would give detailed steps in each of the processes. This information gives us the details about the processes associated with input, output and associative processes. We may need a supplemental input to the information contained in other process documents to finalize requirements but standards and guidelines are usually self-contained.

4. **Customer satisfaction surveys**—These form part of organizational records but deserve special mention as they may bring out inefficiencies inherent in the process. The information contained in these records may be used to enquire if any process changes are necessary before the process is computerized/upgraded.

5. **Customer complaints**—Often, customers of an organization do not wait to voice their complaints till they receive a customer satisfaction survey format. While most of this information could be found in the customer satisfaction surveys, there may still be additional records of customer complaints. These also give us information about the possible improvements in the system which can perhaps be implemented in the proposed system.

6. **Publications/reports/case studies**—on any given subject, we do have many books, best practices reports, experience reports and so on produced by experts/industry associations/consultancy organizations available for study. For example, we now have a host of reports on the implementation experience of ERP (Enterprise Resource Planning) which are very helpful in finalizing requirements of a proposed ERP implementation project. Similarly there are organizations such as Gartner Group, Forrester Research, Technology Evaluation Centers which bring out study reports which are very helpful, for finalizing requirements, especially in product development scenarios.

Thus to reiterate what was stated at the beginning of this section, gathering requirements is collecting information from secondary sources such as documents.

3.4 Elicitation and Gathering in Agile Projects

Agile methodology places more emphasis on satisfying the customer than on documentation. Agile methodology also places emphasis on co-location of the customer with the development team. When the project is executed in-house, this is met easily. But when the project is offshored, this becomes a bit more difficult to achieve. I have seen, what is called "virtual co-location" of the customer. That is, the customer is in his usual location but is available over phone, chat, email and video conferencing to the development team even though both are separated by mighty oceans. And the interesting aspect is that it seems to be working!

Another aspect stipulated by the Agile methodologies is that the iteration is short generally not exceeding four weeks. So, the practitioners of Agile methodologies aver that it is not really necessary to elaborately document the requirements. We may state that the Agile practitioners do not adequately document the requirements but they do capture project requirements, iteration by iteration. They mostly use the elicitation methodology for capturing the project requirements. Still gathering is used, albeit, to a lesser extent especially for report layouts.

How Agile practitioners document their project requirements is a topic covered in the chapter on establishment of project requirements.

3.5 Elicitation and Gathering in COTS Product Implementation

There are many COTS products in the areas of ERP, CRM, SCM (Supply Chain Management), EAI (Enterprise Architecture Integration), Telecom and so on. These products comprise of the best practices culled from their respective industry and cover all functional areas of the selected domain. While so, very few organizations, if at all, can implement the product in total, without having to modify any of their existing business practices. In most cases, some sort of customization would be necessary in the product as well as in the practices. As the products are delivered in the form of executable code, normally a layer would be built over the standard version using the tools provided by the product to suit the practices of the specific organization.

The project requirements for this scenario comprise of the areas of the product which need to be modified for the organization as well as the details thereof for building the layer above the product to meet those needs.

Now, capturing project requirements in this scenario needs special qualifications from the Business Analysts. The Business Analyst ought to be proficient in the product functionality as well as the functional domain.

Both elicitation and gathering would be useful in the projects for implementing the COTS products.

First, if there are any process documents detailing the business practices of the organization are available, they would be studied to capture the gaps between the product and the business practices. Then, product demos would be conducted with the concerned functional executives to see how many gaps need product customization and how many business practices can be modified by the organization. Once this is carried out the project requirements would be ready.

It may sometimes happen that there is little or no useful process documentation available. In such cases, personal interviews and product demos would be the methods to capture the project requirements.

Perhaps, this is the only scenario, in which project requirements are captured by demonstrating the product itself!

3.6 Elicitation and Gathering in Testing Projects

Testing projects differ from other types of partial life cycle projects. In partial life cycle projects, the aim to produce working and defect-free code starts at some point in the software development life cycle. But in the case of testing projects, the software is already developed. The objective of the project could be different for each testing project. The objectives of testing are:

1. To uncover all lurking defects
2. To certify a product as defect-free, virus-free, or malware-free and so on
3. To benchmark a product vis-a-vis other comparable products
4. To accept software from a vendor and start using it

The testing that is carried out concurrently while developing the software is normally referred to as embedded testing. The testing carried out at the end of software development on the final product is normally referred to as product testing. Embedded testing is carried out by the organization which is developing the software. Product testing is sometimes outsourced. Besides unit testing, integration testing, system testing and user acceptance testing there are umpteen variety of tests that are conducted. These include stress testing, load testing, parallel testing, concurrent testing, end to end testing, negative testing, intuitive testing and so on. A testing project could include any combination of those tests. However, when the testing project is initiated, one thing is certain and that is, the code is ready and working. Perhaps, if it was a process-driven project, the requirements and design documents also would also be available.

So what sort of requirements would be needed for a testing project? We can enumerate them below:

1. The type of tests to be conducted as part of the testing project
2. Information for planning the specified tests
3. Information for designing the test cases
4. Information for pass/fail decisions

5. Process for ensuring the quality of the testing process
6. Defect reporting and fixing strategy
7. Regression testing strategy
8. Constraints of time, budget etc.
9. Progress reporting mechanisms

Now, all this information needs to be collected for executing the testing project. This information may be available in project documents or it may have to be elicited from the concerned executives. So, both elicitation and gathering may have to be used in testing projects for capturing the project requirements.

I had seen some testing projects being outsourced without any specifications or requirements or objectives being specified. They simply want the product to be tested and certified. Of course, the testers can point out defects lurking, if any, in the product. All the freedom is given to the testers to choose the type of tests they like to conduct, the method of testing and the test cases design. In such cases, brainstorming also would be a technique to generate project requirements.

3.7 Elicitation and Gathering in Software Maintenance Projects

Software maintenance is said to be consuming about 50 % of a software product life cycle. In some cases, it crossed the 50 % of the life cycle especially in the mainframe COBOL products developed during the sixties and the seventies. Software maintenance work may be carried out in-house or it could be outsourced. When outsourced, the software maintenance project would have an overall contract for the project comprising of the rates, time booking, process for resolution of maintenance work requests, turnaround times, prioritization rules and so on. Then each individual work request would be over the phone, in email or as a formal documented request. Both elicitation and gathering are used in software maintenance projects. When maintenance is carried out in-house, the formal documentation is not rigorously enforced. An informal or barely formal information is given to the IS department which deputes a Business Analyst to elicit requirements. Therefore, elicitation of requirements is used on more occasions than gathering. Some organizations, especially large ones, do use very formal maintenance work requests and in such cases, even in in-house software maintenance, gathering is the most extensively used technique for capturing project requirements. In outsourced projects, the outsourcer and the vendor could be separated by seven seas. Personal elicitation is almost ruled out except for telephonic elicitation. As there is a vendor—vendee relationship involving payments, estimation and approval of estimates comes in. So, formal maintenance work requests are the mostly used form of communication. These maintenance work requests would contain detailed requirements. The vendor needs to ensure that the requirements spelled out are complete and adequate to carry out the work. Therefore, gathering is used more extensively than elicitation.

3.8 Elicitation and Gathering in Real Time Software Development Projects

Real time software, embedded software and firmware are utilized in controlling some sort of hardware. Real time software could run on computers while embedded software and firmware are usually on chips. These projects differ from the projects that produce commercial business process software. For one thing, these software products have very stringent constraints on usage of resources (CPU, RAM, and storage) than in commercial software products. Second, the response time requirements are very critical. The tolerance allowed in the specification of response times is very narrow. Those response times must be met. The requirements come not only from the end user but also the agencies that supply the hardware on which the software resides or functions. Elicitation of requirements from the users as well as the hardware suppliers also need to be carried out. Study of catalogues, product specifications, and component literature is also of paramount importance in capturing the requirements. Sometimes, the products using this kind of software are in fiercely competitive markets. Examples are cars, washing machines, entertainment products like, TVs, DVD players, and set top boxes. In these products, even the cost (money) becomes a constraint and a requirement, which needs to be met by the product. By making the software costlier, the product may lose to competitors in the market as total product cost rises. And missing a requirement could cause the product to lose out in the market on the features front. So, capturing project requirements in these projects is a very delicate balancing affair between capturing comprehensive project requirements and the ability to meet the constraints. Gathering assumes more importance than eliciting in these projects.

3.9 Elicitation or Gathering?

I have come across a misconception especially among software development fraternity that elicitation is the only technique for capturing project requirements. The reality is different. Both are used extensively in the industry. In most scenarios of requirements management, both elicitation and gathering are used in varying degrees. Both are equally important elements of capturing project requirements for the proposed software product. In the case of in-house project development, elicitation is more extensively used to capture project requirements than gathering. In outsourced project development, the requirements are usually supplied to the vendor by the outsourcer. So, the vendor does need not to do much in terms of requirements capturing other than to ensure that the supplied requirements are comprehensive for each of the requirements. In a product development scenario, gathering is utilized much more extensively than elicitation to capture project requirements. It is not a question of elicitation or gathering, it is a situation of elicitation and gathering for capturing project requirements.

3.10 Deliverables of Elicitation and Gathering

What are the deliverables of the elicitation and gathering activities? Perhaps, "deliverables" is not the right word because the end result of these activities would not be in a presentable form. The deliverables of these activities are still in intermediate form needing further transformation. Information is transmitted by the individuals performing business processes to Business Analysts. The following information would be available when elicitation and gathering activities are completed:

1. Notes taken during the personal interviews
2. Responses to questionnaires administered personally
3. Responses to questionnaires administered using the postal method
4. Responses to surveys
5. Formats and templates used in performing the business processes obtained from executives performing those processes. These would also contain all the data items used in the performance of the processes
6. Information culled from studying the organizational records
7. Organizational process documentation, standards and guidelines
8. Flowcharts of process steps
9. List of inputs, outputs and associative processes performed in the organization for the proposed software development
10. Information/notes on how the inputs are converted into outputs
11. Analyses to be carried out on the stored data for preparing the management information
12. Details of required reports for the senior management of the organization.

Of course, all this information would be in raw form, that is, not properly documented/formatted for others to understand and work with. The information can be properly understood and interpreted by the Business Analysts who collated it. When this information is subjected to analysis, it would come to a form that can be used in the down the line activities. We are however, ready for the next activity of requirements analysis.

3.11 Pitfalls in Requirements Elicitation and Gathering

Requirements management is one of the concern areas of the software development. It is also one of the major causes of the failure of the software development projects. Assuming that requirements in the context of software development are understood in their right perspective, the following are the pitfalls in elicitation and gathering of requirements :

1. Untrained personnel being used as Business Analysts is one major pitfall. Originally Systems Analysts (individuals from the software development fraternity who began as programmers and are promoted over a period of time) were capturing project requirements. Presently, Business Analysts who are from functional domains are capturing project requirements. Of course, the breed of Systems Analysts is not extinct. When Systems Analysts capture requirements, design considerations creep in and when Business Analysts capture requirements, budgetary and domain considerations creep in. In the present day, Business Analysts are drawn straight out of business management colleges and put on job. This is one major concern. A new-entrant fresh-out-of-college, even with training on requirements management, would not be able to capture requirements in their entirety. Either a Systems Analyst or a Business Analyst, should have put in a few years of working experience in their respective fields, before imparting training and putting them on requirements elicitation and gathering activity. This is one major pitfall in requirements elicitation and gathering.

2. Bringing in consideration of software design while capturing project requirements—This is another major pitfall especially when the requirements capturing is carried out by Systems Analysts. Requirements capturing is concerned with "what" needs to be achieved and design is concerned with "how" to achieve it. Mixing "how" with "what" causes us to miss some vital information that is essential to fulfill the core functionality of the proposed software product. When capturing the project requirements, we need to take a user's view point. Often times, because of their proximity to software development, the Systems Analysts tend to view the requirements from the software point of view. When this happens, we tend to fit the requirements to the software where as what we ought to be doing is simply capturing the requirements. Whether the captured requirements fit the possible software design is to be considered during requirements analysis stage. We tend to forget that software is proposed to be developed to fulfill the performance of a set of selected business processes but not to give a software product and tell the users to fit their working to the software. This approach of bringing in the design considerations into requirements capturing inhibits the users from giving their view point comprehensively.

3. Bringing in the considerations of time and budgetary concerns during the process of elicitation and gathering of requirements—Business Analysts, especially in COTS product implementation products, take this excuse to

persuade the users to accept the features in the product than to customize the product. How to manage the constraints of time and budget is a case for project management but not for the requirements capturing. This pitfall can be overcome by distinguishing the requirements capturing from project management considerations.

4. Not preparing well when beginning the personal interviews—Personal interviews remain one of the major source of requirements elicitation especially in the project development scenario and is perhaps the only scenario in agile projects. But if we need to elicit information from human beings, preparation is necessary. The users are not well versed in the information requirements of the software developers. Usually, the end users start by narrating the work they do and their concerns and issues. Unless they are carefully and tactfully guided the required information may not be forthcoming from them as they do not know what to give out. So it behooves on the Business Analyst to prepare well before interviewing a person to capture requirements and be ready with right questions to elicit appropriate responses. Asking a wrong question would elicit a wrong and inappropriate response and derails the interview itself. A Business Analyst also has to prepare to ascertain if the user is providing right information or not. If not, the requirements would all be wrong. Business Analysts or whosoever is entrusted with the job of capturing project requirements ought to be trained in the art of interviewing and evoking right response. This sort of training in the industry is rather an exception than a rule. Not preparing adequately before conducting personal interviews is a common pitfall in the elicitation and gathering of project requirements.

5. Prejudices—Both the users and analysts often come with prejudices towards each other's function. Users sometimes show reluctance to part with critical information for the fear of losing their unique indispensability. Analysts sometimes feel that the users do not easily divulge information and it has to be extricated. Sometimes, these are factual and sometimes imaginary. It is essential to instill confidence in users that the proposed computer-based solution is to assist them than to supplant them. This would encourage the users to readily share all the information. It is necessary to train the analysts in the art of conducting personal interviews. That way, this pitfall can be overcome.

6. Omitting the capture of vital process steps and data items in their completeness. Process steps such as validation steps are oft forgotten. The average size of data items is captured instead of maximum size of the data items. Data precision for numeric data items is also oft forgotten. The onus is on the analyst to prod the information sources and obtain complete information. More iterations would be required to obtain complete information if we forget some vital aspects.

3.12 Final Words

Elicitation and gathering is a critical activity in the requirements management. When a project fails due to inefficient engineering and management of requirements, in most cases, the failure would be because of poor requirements elicitation and gathering. In most cases, we carry out elicitation and gathering for the project (or iteration in the case of agile projects) in multiple installments. But the first installment should see that 70–80 % of the requirements being collected. The remaining information may be collected in one or two more installments. If we take more installments for elicitation and gathering, we may often receive contradictory and duplicate requirements. So it is necessary that all efforts must be put in to ensure that this phase is carried out diligently.

Chapter 4
Requirements Analysis

4.1 Introduction to Analysis

When we complete elicitation and gathering activities of requirements management, we have information in the raw form. When we analyze it, we would have plausible requirements which after review and approval would transform into project requirements.

Now what exactly is meant by the term "analysis"?

Merriam Webster's dictionary defines the term "analysis", among others, as:

1. Separation of a whole into its component parts
2. Identification or separation of ingredients of a substance
3. An examination of a complex, its elements and their relations
4. A method in philosophy of resolving complex expressions into simpler or more basic ones

The term "analysis" has a connotation of reducing complexity by breaking down a whole into its component parts to understand how it works and provide better insight into something. Most of us would have come across the following analyses:

1. **Chemical analysis**—In this analysis, a substance is broken down into its component parts. The science of Chemistry has many methods and chemicals to break a substance down to its component parts. A substance that cannot be further broken down except into its atoms is referred to as an element. Chemical analysis is basically breaking down a substance into its elements.
2. **Scientific analysis**—this analysis helps in understanding a phenomenon using scientific method. Scientific method is characterized by controlled and repeatable experiments to test a hypothesis. In scientific experiments, all variables except the ones which are being investigated/evaluated are controlled. The scientific method is also characterized by the absence of bias. That is, the

M. Chemuturi, *Requirements Engineering and Management for Software Development Projects*, DOI: 10.1007/978-1-4614-5377-2_4,
© Springer Science+Business Media New York 2013

same results would be obtained by any person when the experiment is repeated. Scientific analysis helps us in understanding the "why" of a situation.

3. **Statistical analysis**—this analysis aids us in understanding and drawing inferences from a mass of data. Statistical analysis can be performed only on numeric data. Most people are acquainted with statistical analysis, at least to the extent of using averages. It uses measures like measures of central tendency, measures of dispersion, skewness and measures of correlation. It has a number of analysis techniques including correlation analysis, time series analysis, and trend analysis. It has a number of tests such as goodness of fit, hypothesis testing, variance testing and so on. It facilitates determining the probability of occurrence of events. It is one of the most extensively used analyses in physical as well as social sciences.

4. **Mathematical analysis**—using proven mathematical formulas, we analyze the expected behavior of complex systems. One of the easier examples to understand includes the behavior of buildings under various stresses such as earthquakes, tornadoes, and floods. Another is evaluating the behavior of ships and aircraft before building them. It is popularly referred to as "simulation". It is extensively used in product design in all fields including construction, manufacturing as well space exploration.

5. **Medical analysis**—in medical analysis, a patient's condition is analyzed in comparison to a healthy person. The analysis consists of evaluating the symptoms, parameters and diagnostic procedures to aid the medical practitioner to draw inference about the patient's condition. Symptoms act as requirements based on which, further diagnostic procedures are carried out to precisely determine the patient's condition and then using the results, treatment is determined and administered.

6. **Market analysis**—market analysis consists of analyzing the market for a product or an organization. It would analyze the data using statistical methods to ascertain the present market share in terms of the sales and geographical regions. It would also ascertain the relative position of the product or organization in customer perception. Other market analysis using statistical analysis methods include ascertaining the customer needs for a geographical area, demographic analysis, target market analysis, market growth/decline rate, benchmarking of the product or service in terms of pricing and profitability and so on.

7. **Situational analysis**—I think that every one of us carry out this analysis almost every day. Whenever we come across a situation that is not to our liking, we analyze it and draw inferences. But to carry out this analysis properly, we need to have a standard or accepted process. In social situations, we use the standards of behavior to analyze the situation; in religious matters, we use the religious scriptures; in organizational matters, we use organizational process. We compare the situation with the standard situation and draw inferences. These inferences lead to actions.

8. **SWOT analysis**—this is one analysis most of us are familiar with. We enumerate the strengths, weaknesses, opportunities and threats for any given

scenario and draw inferences. While this has origins in the industry, it is now being used extensively in social sciences also.

9. **Business analysis**—Business analysis originated in the IT (Information Technology) field. Originally, it began with comparing an ERP product with the requirements of the business processes of an organization to identify the gaps between both, so that the product could be customized to suit the organization. But, it has expanded beyond the original usage to encompass all the activities in the software development life cycle prior to software design. It now includes the feasibility study to ascertain if a new system is viable, compiling the business requirements, and preparing the requirements specification document. Validation of the end product against the approved requirements has also come under the umbrella of business analysis. While feasibility study and validation of the end product are out of scope for this book, we discuss business analysis role in finalizing requirements in the following sections/chapters.

Summarizing the above discussion, analysis gives us insight into a situation and allows us to draw inferences for taking further steps in the implementation of the solution.

4.2 Analysis of the Information Collected in the Elicitation and Gathering

We carry out the following analysis activities on the information obtained through elicitation and gathering.

1. Enumerate all the requirements
2. Verify each requirement for completeness
3. Evaluate each requirement for its feasibility

    a. Technical
    b. Financial
    c. Timeline

4. Bifurcate requirements into

    a. Core functionality requirements
    b. Ancillary functionality requirements

5. Group core functionality requirements together into logical groups
6. Group ancillary functionality requirements into their logical groups
7. Identify requirements that are duplicated
8. Identify requirements that are contradictory to each other
9. Identify system interfaces
10. Identify stakeholders for each requirement

    a. Primary stakeholder

   b. Secondary stakeholders

11. Prioritize the requirements by the logical group and within every logical group

   a. By timeline
   b. By technical feasibility
   c. By financial feasibility

12. Identify gaps, in the case of COTS product implementation

   a. Between the product and the organizational needs
   b. The needs that can be met by re-engineering the organizational processes
   c. The needs that necessitate product customization

13. Determine the schedule of implementation for the requirements
14. Resolve the issues/inconsistencies uncovered in the above activities

Before we embark on the analysis of requirements, it is essential that we have collected as many requirements as possible from all possible sources. When we analyze the requirements, we may find gaps and may have to go back to sources of requirements once again to fill the gaps. But we need to have almost all the requirements to enable us to analyze the collected information. The tendency is to commence analysis as soon as we collect the information from the end users. Then we would be having only the core functionality requirements. We need to obtain the ancillary functionality requirements too before the analysis can begin. Otherwise, there is every possibility that ancillary requirements are forgotten altogether or added during software design as an afterthought.

Now let us discuss each of these analysis actions in detail.

Enumerate All the Requirements—It is possible that the requirements are elicited or gathered over a period of time either by a single individual or multiple individuals. The requirements are in the form of notes either on paper or on a laptop. The first step in analyzing the requirements is to transcribe them at one place from the raw requirements. This will enable us to perform all the analysis actions enumerated above. It is better to use a requirements analysis tool or a spreadsheet as these will allow us to carry out data manipulation actions such as grouping, and uncovering duplicates. The suggested format is shown in Table 4.1. It is advantageous to use a structured language while enumerating the requirements. That way, we can easily discover duplicates just by sorting the list. We normally begin recording a requirement with a verb followed by two or three more (or a limited number) words. Examples are:

1. Capture login information
2. Raise Purchase Order
3. Raise Invoice

Verify each Requirement for Completeness—We need to verify that every captured requirement is complete. Otherwise, we may not be able to properly classify the requirement or prioritize it. It is also not possible to carry out software

Table 4.1 Suggested format for enumerating all the requirements

Requirement Id	Requirement Description	Functionality (Core/Ancillary)	Logical Group	Stakeholders	Inputs (Y/N/Incomplete)			Outputs (Y/N/Incomplete)		Are Process Steps defined?	Feasibility (Y/N)			Priority
					Defined?	Formats/Sample Documents available	Data Validation defined	Defined?	Formats/Sample Documents available?		Technical	Financial	Timeline	

design from incomplete requirements. To ensure the completeness of a requirement, we ensure that its:

1. Inputs are defined
2. The validations that need to be performed on the input data are defined
3. The process of converting the inputs to outputs is defined
4. The outputs expected from the process are defined
5. All the relevant templates and formats are available

Once we have all this information for a requirement, it can be deemed to be complete.

Evaluate each Requirement for Its Feasibility—Technical—Technical feasibility includes limitations of hardware, software or algorithmic. Sometimes the requirements may not be feasible to be achieved with the current state of technology or the technology available within the organization. Examples include certain types of analyses that are easily performed by the individuals but cannot be performed automatically. Response times are examples for technology limitations. A frequently asked requirement to provide a mechanism to define a new billing plan in consumer services such as cable and mobile phone services is an example of algorithmic limitations. Pattern recognition is also not easily achieved by the software.

Evaluate each Requirement for Its Feasibility—Financial—Sometimes, the requirements may be technically feasible but in our considered opinion, are too costly to fit into the available budget. Such possibilities occur when a specialized hardware or third party software tools are needed to meet the requirement. We have to evaluate each requirement for its financial feasibility.

Evaluate each Requirement for Its Feasibility—Timeline—This limitation is encountered frequently in the industry especially in project scenarios. The requirement is feasible both on a technology basis as well as a financial basis but the timeline cannot be met. It happens because sometimes, the amount of work to fulfill the requirement takes longer than the required timeline.

If there are any requirements that are not feasible due to technical, financial or timeline, we need to resolve them with the end user department. The possible resolutions are:

1. Drop the requirement altogether
2. Postpone the requirement to a future date
3. Increase the budget (financial as well as timeline) to meet the requirement
4. Obtain the technology from outside the organization (if it is available) to meet the requirement

Bifurcate Requirements into Core and Ancillary Functionality—This bifurcation allows us to achieve better grouping of the requirements. This grouping would help in setting priorities as well as during software design. We assess each of the requirements and fit them either as a core functionality requirement or an ancillary functionality requirement.

Group Core Functionality Requirements Together into Logical Groups— We need to group core functionality requirements by the logical group to which they belong. This would help in software design. This can be achieved by taking help from the organization of the function which is the focus of our study. We can take a bottom-up approach here. First we allocate the requirements to the workstation at which it is being performed. Normally a set of operations would be performed at each workstation. Then, we can allocate the requirement to the department/section in which the person operating the workstation reports to. Normally the hierarchical levels from the lowest to top level for a major function would be three or four although exceptions can be found in the industry. The person holding the workstation would report to a section supervisor who would be supervising a set of similar workstations and reporting to a manager. The manager would be managing a few supervisors and reporting to the head of that department. As an example, if we take a warehouse in a supply chain/material management application, there would be a few workstations for material issue and a few workstations for material receipt. Normally all material issue persons would report to a supervisor and all material receipt persons would be reporting to another supervisor. Both of these supervisors would be reporting to the person holding charge of the warehouse. Thus, when we group requirements, we group them into:

1. Warehouse—material issues
2. Warehouse—material receipts

So, we group requirements based on the workstations and departments. This grouping would facilitate grouping of related functions into modules during the software design phase. We need to carry out this assessment for each of the requirements until all requirements are neatly grouped into their logical groups.

Group Ancillary Functionality Requirements into Their Logical Groups— The ancillary functionality requirements are enumerated in Chap. 2. While all core functionality requirements can be designed by one class of designers, all ancillary functionality requirements cannot be designed by one class of designers. Some of the ancillary functionality requirements such as security requirements, and usability requirements would need a different class of designers. Therefore, it is necessary to group ancillary functionality requirements into their logical groups to facilitate better software design.

Identify Requirements that are Duplicated—Especially in large projects wherein requirements are collected from multiple agencies, it is possible that stated requirements are duplicated. We need to eliminate the duplicated requirements so that design effort is not wasted. This can be easily achieved if we had followed requirements description conventions suggested in an earlier section. We can sort the requirements by the "Requirement Description", second column of Table 4.1. We can examine if any requirements are duplicated and if any such duplication is uncovered, we can easily eliminate such duplication. This step is in fact a cleansing step that provides us unique requirements.

Identify Requirements that are Contradictory to each Other—This step is also a cleansing step but is not as easy as locating duplicate requirements. If we have contradictory requirements and allow them to slip through design and construction, we would have a software product that provides unpredictable/unreliable results. To identify contradictory requirements, we need to study each of the requirements; understand it in the right perspective and then see if any other requirement is contradicting the one at hand. The following tips would help in identifying contradictory requirements:

1. When requirements for the same function are collected from two or more workstations, we are likely to receive contradictory requirements. When looking for contradictory requirements, we need to look into the requirements specified for the same function by different individuals.
2. It is likely to have contradiction between the perceptions of the person performing a function and the individuals providing inputs or receiving outputs from that function. So, we need to look closely at the requirements specified by the individual performing a function and the persons providing inputs to that function or receiving outputs from that function.
3. There is also a possibility of contradiction in the perceptions of the person performing a function and his supervisor. It is more likely to be so if the length of experience of these two individuals varies significantly. That is, one of them is new to the function and the other is much more senior. To identify contradictory requirements, we also need to look closely at the information provided by the person performing the function as well as the person to whom he/she is reporting to.

4. There is also a likelihood of contradiction in the perceptions of senior management functionaries and the working level functionaries. So we also need to examine closely the information provided by the senior management personnel and the working level personnel to identify the contradictory requirements.

Summarizing, there is a possibility of contradiction in the requirement if more than one individual provides information for that requirement. So examine all such information and eliminate contradictory requirements,

Identify System Interfaces—End users are likely to cross system boundaries and give requirements that form part of another system. For example, a supply chain system has close linkages with operations as well as finance. End users are likely to give requirements that actually fall within either operations or finance. Similarly, CRM has linkages with operations and finance. The operations department would have linkages with supply chain, CRM, finance as well as human resource systems. Whenever end users specify requirements that actually form part of another system, we need to identify them as requirements for interfaces with other systems. We need to identify all such requirements and classify them under system interface requirements. When we identify a system interface requirement, we need to ensure that:

1. The other system with which the interface is needed is identified
2. The data to be received from the other system is defined
3. The data that is to be transmitted to the other system is defined
4. The protocols, if any, for the interface are identified
5. The entry and exit criteria for the interface are defined

Identify Stakeholders for each Requirement—We need to identify the stakeholders for each of the requirements. These are the persons who need to be approached during the project execution for any clarification about the requirement. The primary stakeholders of course, are the individuals performing the function and the secondary stakeholders are the superiors, providers of inputs to and recipients of outputs from the person performing the function. We need to carry out this task for each of the requirements.

Prioritize the Requirements—Prioritization is the setting of the order of implementing the requirements when there is a resource crunch. If resources are available all requirements would be implemented concurrently. This priority will help the project management to implement requirements of higher priority first if they find themselves short of the requisite resources. Here are some rules for setting priority to help you:

1. **Dominant factor first**—sometimes, it may be possible that a certain function needs to be implemented first. For example, in most applications there would be master data files and without these being created/maintained, the application would not run. Therefore, they need to be implemented first. This type of constraint is referred to as the dominant factor.
2. **Most linked first**—in many applications there would be some modules which would have maximum linkages with the remaining modules. For example, the

purchase order module in a supply chain project or customer relationship management project or account heads in a financial accounting project are modules which would impact many other modules. So these would be implemented first.

3. **First come first served**—we implement the requirements in the chronological order they are received/approved.

4. **Quickest (or the smallest) first**—we implement the requirement that takes the least amount of effort first and the order will be in the amount of effort needed to implement it. The reverse (maximum effort or the largest requirement first) is also possible and is followed in some cases.

5. **Most urgent first**—the order of implementation would be based on the urgency of need specified by the organization where the software would be implemented

6. **Highest benefit**—the order of implementation would be based on the benefit yield by implementing the requirement. The first one to be implemented is the one which would yield the highest benefit and the order of implementation would be based on the decreasing amount of benefit expected from the requirement.

7. **Lowest cost**—the order of implementation would be based on the increasing cost of implementing the requirement. The lowest cost one would be implemented first and the rest would be implemented in the increasing order of the cost. The reverse (highest cost first) is also possible and is followed in some cases.

8. **Tardiest first**—the requirement that is waiting for the longest period is first implemented. This is resorted to when there are many requirements of equal priority and are awaiting implementation.

It is also possible to use a combination of these rules to set priorities for the implementation of the requirements. We need to select the set of rules for setting the priority and then set the priorities for all the requirements.

Normally we set two-level priorities although three level priorities are also used. The first level priority indicates the general priority of implementation. If there is a need to prioritize implementation even within the requirements having identical priority, the second-level priorities are used. For example the first level of priority is based on the urgency. If there are multiple requirements of equal urgency, then we prioritize such requirements based on the amount of benefit they yield to the organization. Third level, if used, would set the priority if the resource crunch is such that even the requirements with second level priority need to be further prioritized.

The resource crunch could be in terms of finances, timelines or technical limitations.

Identify Gaps, in the Case of COTS Product Implementation—Gaps in this context should be understood as the mismatch between the features of the product and the practices/process requirements of the organization. This action is at the core of requirements analysis in the projects that implement a COTS product for applications such as ERP, SCM, CRM etc. During elicitation/gathering phases, we

Table 4.2 Gap analysis

Gap Id	Gap Description	Module Id	Component ID (screen/report/other)	Stakeholders	Bridging action (customize process / product)	Are Process Steps defined?	Feasbility (Y/N)			Priority
							Technical	Financial	Timeline	

would have discussed the product features and the possibility of meeting their needs with the existing product features and collated their views. Now, based on those discussions and the notes taken, we now have to identify the gaps between the product features and the stated organizational needs. We enumerate these differences for all the proposed modules. Using this list, we identify those gaps that need customization of the product. Table 4.2 shows a suggested format for gap identification.

Determine the Schedule of Implementation of Requirements—Scheduling is the activity of assigning calendar dates to planned activities. When we carry out requirements analysis, we may not be in a position to assign absolute calendar dates to implement the requirements because we may still need to resolve some of the issues uncovered during analysis itself. Based on the priorities assigned to all the enumerated requirements, we can prepare a tentative schedule with an assumed start date. If we use a tool such as MS Project, Primavera or PMPal, we can shift the start date and the tool would take care of the rescheduling of the remaining dates. Before we schedule the implementation of requirements, we need to estimate the effort required for the remaining activities of software development and implementation (if included). This schedule would assist the end users to assess the project execution and plan their further activities. Discussion on effort estimation and scheduling is beyond the scope of this book. Interested readers are advised to refer to the book "Software Estimation: Best Practices, Tools and Techniques for Software Project Estimators" by Murali Chemuturi, Published by J.Ross Publishing, Inc. 2009.

4.3 Resolving the Issues that Cropped up During the Analysis

We have noted during the discussion on analyzing the requirements that some issues/inconsistencies could crop up. These issues could be stemming from requirement feasibility, shortfall of information for inputs, outputs or process steps, contradictory requirements, duplicate requirements, logical grouping of requirements, and prioritizing the requirements. We need to resolve all the issues before we move on to the next phase of requirements management. We need to, perhaps, go back to the end users or their superiors or technical or domain experts or whosoever provided us the information and discuss the issue and resolve all the issues. Once we resolve an issue, we need to update the enumerated requirements to reflect the resolution.

4.4 Deliverables of Requirements Analysis Phase

Upon completion of the requirements analysis, we would have the following

1. A list of all requirements
2. A list of gaps, in the case of COTS product implementation
3. All issues in requirements are resolved
4. Priorities and a tentative schedule for implementation of requirements
5. Stakeholders for all requirements are identified
6. All requirements are logically grouped

We would only have one document, perhaps a spreadsheet or information inside a tool that would have all the information noted above. We would also have the information collated in the elicitation and gathering phase. Now we are fully armed to begin the establishment of project requirements.

4.5 Final Words

It is normal practice in the industry to carry out this phase informally. That is no formal enumeration of requirements is carried out. All the steps in the analysis detailed in the foregoing sections are also omitted. What normally happens is that these activities are combined with the establishment of requirements. Once the requirements information is collated during the elicitation and gathering phase, the preparation of the requirements specification document is embarked upon and the analysis steps discussed in this chapter are concurrently carried out. In smaller projects of short duration, perhaps, analysis can be combined with establishment of the requirements. But in larger projects and especially in product development, it pays to formally carry out analysis. I have seen technical failures of large projects and products stemming from poor requirements management and the requirements management failed because the requirements were not analyzed properly in the first place.

Most people who use the term "analysis" understand it poorly. I have seen this poor understanding even in some people who hold the title of "Business Analyst" in organizations. That is the reason why I included information about various connotations of the term right at the beginning of the chapter to clear the air about it and give the reader a proper grounding in the art of requirements analysis. This should help individuals to understand requirements analysis in its right perspective and pave the way for better requirements management in software projects and thus reduce the technical failure of projects due to poor requirements management.

Chapter 5
Establishment of Requirements

5.1 Introduction to Establishment of Requirements

Merriam Webster's dictionary defines the term "Establishment" as "a settled arrangement", "as established order of society" among other definitions. CMMI® model document for development used the term "establishment" in the context of requirements but left it without a definition. The term "establishment" is defined here on the lines of the Merriam Webster's dictionary. That is, *"Establishment of requirements in a software project is defined as the documentation of the project requirements conforming to organizational documentation guidelines, carrying out applicable quality assurance activities, obtaining the required approvals and subjecting the document to the rigor of organizational configuration management."*
The definition includes four key aspects, namely,

1. **Documentation**—It is the act of capturing all the information compiled as requirements in a structured manner conforming to organizational/project documentation guidelines and formats/templates. All compiled information that is subjected to the requirements analysis, is included in the document. Each requirement has all the details necessary to carry out the next activity. The resulting document is generally referred to as the Requirements Specification" or "Requirements Specification Document".
2. **Quality Assurance activities**—These activities ensure that quality has been built into the artifacts. Some of these activities are at an organizational level to create an environment that fosters quality in all the work carried out in the organization. The others are at project level which create project specific environments for building quality into the artifact and to confirm that quality is indeed built into the artifact. In the case of requirements management, the environment at an organizational level includes training, processes, standards, guidelines, formats and templates etc. and the environment at the project level includes project plans, and project specific training, standards, guidelines,

M. Chemuturi, *Requirements Engineering and Management for Software Development Projects*, DOI: 10.1007/978-1-4614-5377-2_5,
© Springer Science+Business Media New York 2013

formats and templates etc. The activities that ensure that quality is indeed built into the artifact are verification and validation.

3. **Obtaining approvals**—An artifact is not deemed fit for use in the project until it is approved by the concerned executives. In the case of Requirements Specification, the approvals are accorded by the IS department (in the case of internal projects) or the concerned project manager (in the case of external projects) as well as the approvals of the end user department or the client organization. These two levels of approvals authenticate the information contained within the artifact for use on the project.

4. **Configuration management**—Configuration management is the set of activities that ensure that any changes to approved artifacts are affected in a controlled manner. That is each change request is raised, evaluated, impact ascertained, approved/rejected and approved change requests are implemented in the artifact conforming to an organizational process and the project plans. Covering configuration management comprehensively is beyond the scope of this book and interested readers are referred to the book "Mastering Software Project Management: Best Practices, Tools and Techniques".[1]

Performing all the above four activities diligently is referred to as the "establishment of requirements" for the project. Let us discuss all these activities in detail in the following sections.

5.2 Documentation

As noted in Chap. 2, we have two requirements specifications. One is User Requirements(URS) Specification aimed at capturing the needs of the end users and the other is Software Requirements Specification (SRS) aimed at guiding the design and construction of the product. IEEE terms these two documents as Systems Requirements Specification (SyRS) and Software Requirements Specification (SRS). However, individual organizations may have different names for these two documents as noted in Chap. 2. We will discuss both these documents in this section.

Documenting the requirements, be it URS or SRS, is describing each of the captured requirements in detail with all relevant details in a comprehensive and structured manner so that the concerned stakeholders can understand the requirements without any ambiguity and ensure that their stakes are well provided for. The stakes for the providers of information for the document are that their requirements are understood as intended and accurately captured by the software development team. The stakes for users of the document are to ensure that all the necessary information is available in the documents to carry out their own downstream activities. The crux of documenting the requirements lies in ensuring that there is no ambiguity in the documented requirements. Normal free-flowing

---

[1] By Murali Chemuturi and Thomas M. Cagley, Jr., published by J. Ross Publishing, Inc., 2010.

language is prone to inject ambiguity into the document, especially in the English language, which is unusually rich in the availability of multiple words with similar (not identical) meaning. It is also rich in the adages, idioms and phrases which are not uniformly interpreted all-over the world. Regional flavors influence interpretation of the meanings of the written word. Again, English language also facilitates obfuscation of meaning very easily. Therefore, it is common practice to restrict the free usage of language in documenting the requirements in the software development organizations.

The normal practice in organizations is to eliminate ambiguity in the documents, to define a set of documentation guidelines and to ensure conformance to that guideline in all software engineering documents. All standards organizations such as IEEE, ANSI, JSS as well as government departments do have their own documentation guidelines for preparing professional documents. A suggested documentation guideline is provided in Appendix A. Another alternative is the Planguage of Tom Gilb[2] which is briefly described in Appendix B. IEEE is working on this aspect in their project P1805—*Guide for Requirements Capture Language* (RCL). We can also use those guidelines in our organization with or without change. However, every organization is unique in its own way and it is better to derive an organizational documentation guidelines/standard appropriate for its unique environment and use it in the documentation of requirements. This organizational guideline can draw upon the best aspects of all such available guidelines.

One other aspect worth discussing here is the methodologies available for documenting the requirements. There are literally a plethora of methodologies continuously coming into, and going out of, fashion in the software engineering field. These include SSADM, OOM, UML, and Agile methodologies (Scrum, DSDM, Clearcase, XP and so on). These are all in vogue in different organizations and each has its own followers. However, those methodologies are used to describe the heart of the document. The other sections of the document such as title page, table of contents, system information, constraints, and closing pages are left uncovered by these methodologies. Therefore, in the templates provided in the following sections, I have indicated where specific methodological deviations can occur.

5.2.1 User Requirements Specification

URS captures all the requirements provided by the end users as well as other sources at one place in a comprehensive and structured manner. This document would be used by two sets of audiences, namely the end users and other providers of requirements as well as the software developers. Providers of requirements would refer to this document to ensure that their requirements are understood as

---

[2] Book—Competitive Engineering: A handbook for Systems Engineering, Requirements Engineering , and Software Engineering using Planguage , Elsevier Butterworth, Heinmann, 2005.

intended and captured accurately by the business analysts properly. The end users would use this document as the reference point for all interactions with the software development team and in accepting the final product when it is delivered. The software development team would refer to this document to carry out the subsequent software development activities namely derivation of product specification, design of the proposed software product/system, and developing the user acceptance test plan of the product. We need to document URS in such a manner that the users would have no trouble in understanding the document and accord approval to the document. To ensure that aspect, we need to use the user's language while documenting the URS. Or, in other words, we need to avoid any jargon normally part of software development fraternity or the software development organizations. IEEE standard 1233 deals with defining SyRS. It is common for organizations to have in place documentation guidelines to ensure that free-form language that can give rise to ambiguities, vagueness or conflicting meaning, has not crept into the document. A suggested set of documentation guidelines are provided as Appendix A to this book.

We need to document the requirements in the logical groups we arranged while analyzing the requirements. It is also advantageous to use the same nomenclature that is in the list of requirements generated during the requirements analysis. That will maintain consistency between the two documents. A suggested template for documenting URS is given in Table 5.1. It contains essentially five sections: Title page, table of contents page, project information page, requirements definition pages and closing pages including appendixes, if any.

In agile software development methodology projects, it is common to replace URS with a set of *User Stories* which describe the requirements in a descriptive manner. In true conformance to the agile philosophy, no template or format is commonly advocated and the stories are formatted differently in different organizations. The cardinal rule in this case is that the project team, which usually comprises the client representative, must be comfortable in the format and be able to use it to develop the software. Therefore, no attempt is made here to present a format for documenting the User Stories in this book.

So is the case with Use Case methodology. Use Cases are also used to document URS. A *Scenario* description is used to capture user requirements along with the Use Case diagram. There is no standard format or template commonly advocated in Use Case methodology to document the requirements.

5.2.2 Software Requirements Specification

A Software Requirements Specification (SRS) document may be referred by different names in individual organizations as explained in Chap. 2 but, irrespective of the specific name used in the organization, the document contains the specifications for the proposed software product/system. I am using the name "SRS (Software Requirements Specification)" in this book because it is christened so by

Table 5.1 Suggested template for documenting URS

<Title Page>			

User Requirements Specification
For *<Project Name>*

Revision History:

Version Number	Description of Changes	Date	Approved by

<end of title page>

<Table of contents page>

<Insert table of contents on this page>

<End of table of contents page>

<Project information page. It can span across multiple pages.>
Basic Information about the Project

Project Id :

Project Name :

Business Processes covered by this URS :

Business Process	Primary Stakeholder	Other Stakeholders

References :
<Give a list of all the documents referred in the preparation of this document. This list may include client supplied documents, standards, guidelines, formats and templates.>

    1.
    2.
    3.

Scope of the system:
<Describe the scope of the proposed software product/system in this section. Define the boundaries of the proposed system, the activities included, and the activities excluded from the purview of the proposed system>

(continued)

Table 5.1 (continued)

Critical Success Factors:
<Describe the conditions that facilitate determination if the proposed software product / system is successful from the end-user perspective>

   1.
   2.
   3.

Constraints specified by the end-user:
<Describe all the constraints specified by the end-users or management of client organization. They may include delivery date, hardware specifications / limitations, system software specifications / limitations, applicable statutes, preferred colors, graphics usage, logo positioning on user interface or any other specifications or limitations placed on the proposed software product / system>

   1.
   2.
   3.

<Requirements definition pages>
<Requirements are likely to span across multiple pages. Therefore, it is a better practice to begin every requirement on a new page. Here is a template for one requirement and it can be replicated for each of the requirements. This section should include all requirements including core functionality requirements as well as ancillary functionality requirements. Agile methodologies utilize User Stories to describe the requirements. Use Case methodology uses scenarios to describe the requirements.>

1. **Module Id**<*The logical grouping to which this specific requirement belongs to may be mentioned here.*>
2. **Requirement Id**<*Use the same id generated during requirements analysis as shown in Exhibit 4.1 in chapter 4*>
3. **Stakeholders**<*Enumerate all the stakeholders for this requirement*>
4. **Priority**<*indicate the priority in which this requirement needs to be implemented*>
5. **Requirement Description**<*describe the requirement in detail in this section.*>
6. **Inputs**<*Enumerate all the inputs here. Ensure that each of the inputs is defined along with its data type, source from where it is received, width necessary to accommodate the data, and its validation criteria and associated rules.*>
7. **Outputs**<*Enumerate all the outputs here. Ensure that each of the outputs is defined along with its data type, destination (screen/printer/other hardware) and width necessary to accommodate the data.*>

(continued)

Table 5.1 (continued)

8. **Process**<*Describe the process of receiving inputs, store the processed inputs, receive parameters for generating the outputs, the process of converting inputs/stored information into outputs, protocols of transmitting the outputs to the output media in this section.Use diagrams where necessary. The diagrams may be flow charts, DFDs, Use Cases or any other applicable diagrams approved for use within the organization*>   9. **Acceptance criteria**<*Describe how this requirement would be tested by the client to ensure that the product meets specifications. Describe the test cases if possible.*>   <*End of requirements pages*>   <*Closing pages and appendixes, if any*>   <*This section may contain acknowledgements, list of executives that provided the information, the list of documents used in gathering requirements, list of statutes referred to and so on.*>

IEEE. If you expand the acronym SRS as "Software-Requirements Specification" (combining the first two words), the meaning would become clearer. It is the specification document detailing the requirements for the software product/system from the technical perspective. Perhaps, it would be even more lucid if we term it as "Software Product Specification", but it would be adding another term to the already bulging vocabulary of software engineering.

SRS is derived from the requirements specified by the end-users in the URS. While URS documents requirements from the perspective of the end-users, SRS documents requirements from the perspective of the product. SRS is the statement of technical specifications for the proposed software product. SRS is the main reference for the product designers in carrying out the design of the proposed software product/system along with other organizational standards and guidelines for software design.

Prototype is often used to supplant SRS in some organizations. Those organizations document user requirements in the URS and build a prototype instead of deriving and documenting the software requirements. The main advantage of doing so is that the end-users can better perceive the proposed software product/system when presented with a prototype rather than a SRS document. The downside of building a prototype is that much more effort needs to be spent in building a prototype than in documenting SRS. Second, if the user-requirements are poorly understood, the prototype may have to undergo major changes and it would consume significant effort in overhauling the prototype than it would take in overhauling a SRS document.

Prototype is used heavily in product development in the manufacturing industry. When it is contemplated to build a number of products with the same design, such as automobiles, it is common practice to build one car with the design and subject it to all necessary tests. When the prototype successfully passes through all the tests, the mass manufacture begins. In these cases, the prototype is used more to prove the product design than to document the product specifications.

Prototypes are also used when building one-of-a-kind product such as ships, aircrafts, rockets and so on. The quantity of these products would be just one. So building a full scale prototype is not possible. In these cases, the prototype is a scaled down model of the proposed product. The prototype is subjected to all applicable tests in the laboratory. Once the prototype passes through all the tests, the main product would be built. Here too, the prototype is built mainly to prove the design than to document product specifications.

Thus as you can see, prototype is not really used to document product specifications in the manufacturing industry but to prove the design. The step of documenting product specifications is not skipped in the manufacturing industry. The design for the proposed product must be ready to begin building prototype.

But in software industry, in some organizations, prototypes are advocated to supplant the product specification document. The prototype would contain the fulfillment of both the user requirements and product specifications but it is not possible to move backwards from prototype towards extracting the product specifications.

In agile methodologies, user stories are used to carry out software design. They skip the step of documenting the product specifications. Agile goes from user stories direct to design and then on to product construction. In most cases of agile development, they go straight from user stories to product construction skipping both product specification and product design. Agile adherents consider effort spent on documentation other than which is absolutely essential as wasted effort.

Software development is unlike manufacturing as it does not involve the making and breaking of any physical material. A mistake is not visible to everyone; it is not lying on the floor in front of everyone but lurking inside the computer; one needs to open the program to locate it. Correcting a mistake in manufacturing involves a lot of noise made by the equipment, and waste of material. Correcting a mistake in software development is just "debugging" without involving noise or material wastage. On the other hand, to determine the material defect, one needs to carry out chemical analysis in manufacturing. In software development, looking at source code is adequate to understand the defect.

So, what are the conclusions? Do I advocate preparing an SRS or skipping it?

When it comes to building a prototype, I am not very enthusiastic. We need to take approval from the customer/end-users only for the URS. SRS is inherently an internal document for the guidance of the software designers. Prototype is essential to prove the product design. But I have had occasion to see wherein the end-user organization is not able to decipher even the URS. They want some solution to their problems in information management but expect the software developer to use his/her expertise to come up with the solution for the problem that is not properly defined by the end-users. In such cases, even the URS is not approved. In such cases, a prototype is perhaps the only solution to move forward, wherein a prototype is URS, SRS and design all rolled into one. Except in such extreme cases, I do not see or recommend the use of prototype to supplant SRS.

Coming to the practice of combining URS and SRS into one document like the user story or the scenario, we need to take the specific instance into cognizance. Iterative construction of the product is possible in many cases but not all. Building one thousand row houses is possible to achieve in ten iterations of hundred houses per iteration. But constructing a building like the Empire State Building, or the Sears Towers or the Petronas Towers or the Burj Dubai is not workable in iterations. So if the proposed software product/system is amenable for iterative construction, perhaps SRS may be skipped. Another important consideration is the need for software maintenance to determine the need for an SRS. If the development language is of self-documenting variety or has an IDE (Interactive Development Environment) that makes it easier to understand the source code and maintain it easily, perhaps, we can skip SRS.

Barring those specific cases mentioned in the above paragraph, I humbly submit that skipping the SRS is not a good practice. There are standards of "GMP" (Good Manufacturing Practices) for various types of manufacturing industry. I wish that our software development industry can also come up with "GSDP" (Good Software Development Practices), some day!

Table 5.2 provides a suggested template for documenting SRS. It has essentially eight sections, namely, the title page, table of contents, project information, hardware and system software specifications page, interface specifications page, core functionality specification pages, ancillary functionality specification pages, and the closing pages.

5.3 Quality Control of the Documents

In the establishment of requirements, ensuring that quality is built into the requirements specifications is very important. The impact of an error that escaped being caught and rectified in this phase would have serious impact when uncovered in later stages. If the escaped error is caught in software design phase, we have to retrace our steps by one level before we can move forward. If the error is caught in the software construction phase, we need to retrace our steps by two levels and if the error is caught during the acceptance stage, we need to retrace our steps by three levels besides the embarrassment of being pointed out by the customer. If the error is caught after the software product/system is implemented, we might even have to suffer losses. The later the error is unearthed, the greater is its impact and loss. Therefore, we need to subject these requirements specification documents to diligent quality control activities. The possible quality control activities are verification and validation. Quality control and quality assurance activities are very important in software development. Therefore, we have dedicated a separate chapter for a detailed discussion of these activities including concepts. These are discussed in greater detail in Chap. 6.

Table 5.2 Suggested template for documenting SRS

<Title Page>
Software Requirements Specification
For *<Project Name>*
Revision History:

Version Number	Description of Changes	Date	Approved by

<end of title page>

<Table of contents page>

<Insert table of contents on this page>

<End of table of contents page>

<Project information page. It can span across multiple pages.>
Basic Information about the Project

Project Id :

Project Name :
Project Manager:
SRS Prepared by: Date:

References :
<Give a list of all the documents referred in the preparation of this document. This list may include client supplied documents, standards, guidelines, formats and templates. Each document reference should include the title of the document, its version number, its source, and if obtained from a published source, the name of the publisher and the date of publication.>

1.
2.
3.

Scope:
1. Name of the proposed product / system:
2. Functional Domain of the proposed product / system:
3. Functionality of the proposed product / system:
4. Objectives of developing the proposed product / system:

(continued)

Table 5.2 (continued)

Overview of SRS:
<Describe the SRS in brief including its contents and organization.>

<Hardware and software specifications page>
Hardware and System Software Specifications:
<It is possible that this specification may be postponed to design stage in some cases. Please fill this if this information is available. Otherwise, indicate that this decision is postponed to design phase.>

Machine	RAM capacity	HDD Capacity	OS and other system software	Any special hardware
Web Server				
Database Server				
Application Server				
Client Machines				
Any other hardware				

Additional Software Specifications:
<Specify the bowsers to be supported, the HTML / XML versions to be supported, and any other such special software requirements in this section.>

<Interface specification pages>
Interfaces with external systems :
<Describe the interfaces needed by the proposed product / system with other systems excluding human interfaces>

Name of the external system	Communication protocol	Data expected to be received	Data expected to be transmitted	Triggers for interface activation

Constraints: *<Some examples are shown for illustrative purposes in the below table>*

Name of the Constraint	Reason	Maximum capacity allowed	Minimum Required	Ant other information
RAM available for the software	User selected hardware	32 MB		

(continued)

Table 5.2 (continued)

Storage	User selected hardware	1 GB		
Internet speed	User environment	1MBPS		
Response time	User selected hardware	30 milli-seconds	15 milli-seconds	
Day-end processing	User requirement	Should trigger at 5 PM on all days		
Week-end processing	User requirement	Should trigger at 5 AM on Monday morning		
Monthly processing	User requirement	Should trigger at midnight after the last working day of the month		
Yearly processing	User requirement	Should trigger on the midnight of 31st December		
Backup process	User requirement	Should trigger immediately after day-end processing		

General Product Specifications:

<Product specifications for general requirements: these cover specifications that are to be considered in general for all requirements. You need to include guidelines for user interface (both screen layouts and report layouts), design, common data validation guidelines, special usability guidelines etc.>

1.
2.
3.

Core Functionality Specifications :

<Product specification pages for core functionality requirements: These are likely to span across multiple pages. Therefore, it is a better practice to begin every specification on a new page. Here is a template for one specification and it can be replicated for each of the specifications. Agile methodologies use User

(continued)

Table 5.2 (continued)

Stories to describe the requirements. Use Case methodology uses scenarios to describe these specifications.>

1. **Module Id<***The logical grouping to which this specific requirement belongs to may be mentioned here.>*
2. **Requirement Id<***Use the same id generated during requirements analysis as shown in Exhibit 4.1 in chapter 4. It is possible that a specification may cater to more than one user requirement. Hence, it may include multiple requirement id's.>*
3. **Specification Description<***describe the product specification in detail in this section.>*
4. **Inputs<***Enumerate all the inputs here. Ensure that each of the inputs is defined along with its data type, source from where it is received, width necessary to accommodate the data and its validation criteria and associated rules. Please note that each input described here may result in one screen layout or a program unit for receiving inputs from an external system.>*
5. **Outputs<***Enumerate all the outputs here. Ensure that each of the outputs is defined along with its data type, destination (screen/printer/other hardware) and width necessary to accommodate the data.Please note that each output described here would result in a report, or an enquiry screen layout, or a program unit to transmit information to an external system.>*
6. **Process<***Describe the process of receiving and validating the inputs, storing the processed inputs, receive parameters for generating the outputs, the process of converting inputs/stored information into outputs, protocols of transmitting the outputs to the output media in this section.Use diagrams where necessary. The diagrams may be flow charts, DFDs, Use Cases or any other applicable diagrams approved for use within the organization>*

Ancillary Functionality Specifications :
   <Product specification pages for ancillary functionality requirements: These are likely to span across multiple pages.>

1. **Statutory Requirements<***Enumerate the statutory aspects which have to be met by the proposed software product / system.>*
2. **Safety Requirements<***Enumerate all the aspects that could cause loss to the users if not taken care of in the design.>*
3. **Security Requirements<***Enumerate all aspects necessary to keep the system from external attacks and snooping.>*
4. **Data Integrity Requirements<** Enumerate all aspects to keep the integrity of system data intact. This is normally a design aspect but we can mention any special requirements over and above normal design considerations.>

(continued)

Table 5.2 (continued)

5. **Response Time Requirements**<*Enumerate in detail the response times required for each of the user actions in quantifiable terms. Avoid terms such as "quick" or "fast" or "immediate". Give numerical values with time units.*>

6. **Fault Tolerance Requirements**<*Enumerate the scenarios where fault tolerance is needed and the workarounds possible for each of the scenarios.*>

7. **Reliability Requirements**<*Enumerate scenarios that could cause reliability issues such as changes in browser versions, malware attacks, installation of other utilities and so on so that design phase could consider the eventuality while carrying out the software design.*>

8. **Competitive Edge Requirements**<*Enumerate all aspects in which the proposed product / system needs to be better than the competing products. Describe in detail the existing best features of competing product and what betterment is desired.*>

9. **Maintainability Requirements**<*Coding guidelines normally ensure maintainability of the code. So, enumerate any special guidelines that are not covered in the organizational coding guidelines.*>

10. **FlexibilityRequirements**<*Enumerate the aspects in which flexibility is desired such as billing plans, tax computations, new classes of data etc.*>

11. **EfficiencyRequirements**<*Enumerate all aspects of efficient resource (memory, disk space, bandwidth, human effort etc.) necessary for the proposed product / system.*>

12. **ReusabilityRequirements**<*Enumerate all those units of the system which are proposed to be reused as well as the level (source code level or executable level) at which reuse is foreseen.*>

13. **PortabilityRequirements**<*Enumerate all those foreseeable requirements to port the proposed product / system to other similar platforms. Also mention the possible platforms to which code porting would be necessitated.*>

14. **TestabilityRequirements**<*Indicate at what level the proposed product / system needs to be testable by persons independent of the project team. That is at unit level or module level or system level.*>

<*Closing pages and appendixes, if any*>
<*This section may contain acknowledgements, list of executives that worked on the SRS, the list of documents used in deriving SRS, list of statutes referred to and so on.*>

5.4 Obtaining Approvals

A document is not frozen until it is approved by an appropriate authority. Approvals are required to be obtained from two sources, namely, the internal sources and the client organization. Normally URS needs both the approvals.

SRS is normally approved by the internal sources unless the contractual arrangements mandate approval from the customer.

Before, the document is submitted for internal approval, it should have been subjected to the planned quality control activities. The approving authority would ensure that the document has indeed passed through the quality control activities and is cleared for approval. Then he/she would carry out a managerial review of the document and provide the feedback to the originator of the document. Managerial reviews are detailed in Chap. 6. Once the feedback is implemented to the satisfaction of the approving authority, internal approval is granted. The internal approval may be a signature on a hard copy of the document or it can be an email or it can be digital signature depending on the practice within the specific organization. The approval information in whatever form, is stored for future reference and records. In the case of requirements documents, in some organizations, the person holding charge of the service delivery department (also frequently referred to as the technical head or delivery head) accords internal approval and in others, the software project manager accords the internal approval. Normally the approving authorities for various documents are named in the organizational process documents.

The artifacts can be submitted to customer organization for approval only after obtaining the internal approval. In the case of internal projects, approval of the IS department is essential before submitting to the client department. The customer, be it internal or external, needs to review the document before according approval to move forward on the project. This review can be a guided review or a postal review. Various types of reviews are explained in Chap. 6. The client feedback, if any, needs to be implemented in the document and re-submitted for approval. The client would review the implementation of the feedback and upon being satisfied that the requirements are properly understood by the software development organization, approval would be accorded to the document.

It is normal practice to submit URS for client approval. SRS would be submitted for client approval only in cases where it is a contractual requirement. Normally, clients would not like to look at SRS as it is more of a design document than a requirements document.

5.5 Configuration Management

Configuration Management is an important aspect of software engineering and software project management in ensuring that the right software artifacts (code artifacts and information artifacts) are delivered to the customer. Configuration management assumes much more importance in software development, than in manufacturing, as the artifacts are maintained in soft copy form, it is possible for different versions of an artifact to exist concurrently. In manufacturing, two versions of the same part cannot exist at the same time. While it is possible for different versions of engineering drawings to exist concurrently, the drawing release process ensures that earlier versions are withdrawn before newer versions are issued.

Therefore software development organizations implement a rigorous configuration management process to ensure that the integrity of project's software artifacts is protected. It is protected using a configuration management tool or a directory/folder structure coupled with strict security enforcement. Once a software artifact is brought under the rigor of configuration management, changes to the artifact are strictly controlled conforming to the organizational change management process. We discuss change management as applicable to requirements management in Chap. 8. When is the right time to bring the requirements specifications documents under the rigor of configuration management? As a general rule, a software artifact is brought under the rigor of configuration management only after the final level of internal approval is accorded to the artifact. So, if we receive any feedback from the customer as a response to our submission for approval, the artifact would still be subjected to change management process.

Establishment of requirements is depicted pictorially in Fig. 5.1.

5.6 Establishment of Requirements in COTS Product Implementation Projects

COTS product implementation products differ from software development projects because the software product is already available. But, it has to be customized, either by building a layer over the product which is the case most of the times or modifying the source code which is the case in a few cases. The stages in implementation of a COTS product from the software engineering standpoint, in an organization are:

1. **Gaps analysis**—This is comparing the functional practices existing in the organization with the functionality available in the COTS product to bring forth the gaps between them. The gaps are recorded in a gaps analysis document. Each of the gaps is analyzed to determine one of the three possible alternative courses of actions to bridge the gap. The three possible alternatives are to customize the COTS product, customize the organizational practice or take no action and leave both as they were. We discuss this document in detail in the subsequent sections
2. **Statement of Work (SOW) preparation**—This draws upon the gaps analysis document to determine the customization necessary for the COTS product. The gaps that would be bridged by customizing the organizational functional practices are omitted in the SOW document. The gaps that need customization of the COTS product would be included in the SOW document. We will discuss this document also in the subsequent sections.
3. **Software design**—Using the SOW document, software design is carried out to bridge the selected gaps. When a layer is proposed to be built over the COTS product, a Software Design Document would be prepared. When it is proposed to modify the source code of the COTS product, a Conversion Guidelines document would be prepared. This activity is beyond the scope of this book.

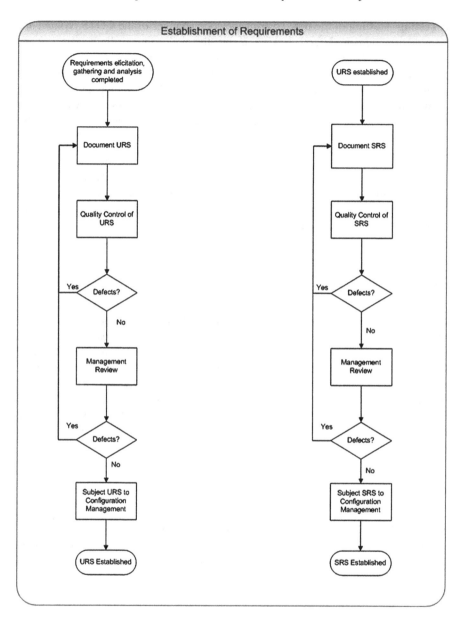

Fig. 5.1 Establishment of requirements

4. **Construction of the code to fulfill the SOW**—Code will be developed or customized conforming to the software design. When building a layer over the COTS product, new source code would be developed. When the source code is modified, the code changes would be implemented in the source code conforming to the Conversion Guidelines document.

5. **Testing**—testing of the new layer or the code changes of the COTS product are tested to uncover defects lurking inside the code. Once all the defects are rectified, the code would be passed on to the next stage.
6. **Implementation**—during this stage, the COTS product would be implemented. Then if a new software layer is built, it would be implemented. Now the implementation is tested to ensure that all functionalities are working as designed without any defects, it would be commissioned for use in the organization.

The above description is very brief and meant just to record an overview of an elaborate process for COTS product implementation. Different organizations would have different stages/phases with different nomenclature. For the purpose of this book and the topic, gaps analysis and SOW documents are relevant and would be discussed in greater detail.

5.6.1 Gaps Analysis Document

This document records the gaps between the COTS product and the functional practices of the client organization. Table 4.2 in Chap. 4 gives the suggested format for capturing the gaps uncovered during the requirements analysis phase of COTS product implementation projects. This is the first level document prepared in the projects implementing a COTS product (such as ERP, SCM, CRM, or a data warehousing product and so on). This document is the equivalent of URS in software development projects. Based on this document, the SOW document would be prepared.

5.6.2 SOW Document

SOW document in COTS product implementation projects may be viewed as the equivalent of the SRS document in software development projects. It documents the product specifications for the COTS product implementation projects.

SOW document would normally have four sections, namely, the title page, table of contents page, SOW pages and the closing pages. A suggested template for capturing the SOW is given in Table 5.3.

Using this document, software design to build the additional layer or modify the code is carried out.

The activities of quality control, obtaining approvals and configuration management are as described in Sects. 5.3, 5.4 and 5.5.

Table 5.3 Template for SOW

<Title Page>	

Statement of Work
For *<Project Name>*

Revision History:

Version Number	Description of Changes	Date	Approved by

<end of title page>

<Table of contents page>

<Insert table of contents on this page>

<End of table of contents page>

<Project information page. It can span across multiple pages.>
Basic Information about the Project

Project Id :

Project Name :
Project Manager:
SOW Prepared by: Date:

References :
<Give a list of all the documents referred in the preparation of this document. This list may include client supplied documents, gaps analysis, standards, guidelines, formats and templates. Each document reference should include the title of the document, its version number, its source, and if obtained from a published source, the name of the publisher and the date of publication.>

   1.
   2.
   3.

<SOW pages>
Statement of Work :
<SOW pages: These are likely to span across multiple pages. Therefore, it is a better practice to begin every work statement on a new page. Here is a template for one statement and it can be replicated for each of the statements.>

(continued)

Table 5.3 (continued)

1. **Module Id**<*The COTS product module ID to which this statement belongs to may be mentioned here.*>
2. **Gap Id**<*Use the same id generated during gaps analysis as shown in Exhibit 4.2 in chapter 4. It is possible that a statement may cater to more than one gap. Hence, it may include multiple gap id's.*>
3. **Specification Description**<*describe the work statement in detail in this section.*>
4. **Inputs**<*Enumerate all the inputs here. Ensure that each of the inputs is defined along with its data type, source from where it is received, width necessary to accommodate the data and its validation criteria and rules. Please note that each input described here may result in one screen layout or a program unit for receiving inputs from an external system.Give reference to the COTS product tables. If new tables are needed, give the table structure also*>
5. **Outputs**<*Enumerate all the outputs here. Ensure that each of the outputs is defined along with its data type, destination (screen/printer/other hardware) and width necessary to accommodate the data.Please note that each output described here would result in a report, or an enquiry screen layout, or a program unit to transmit information to an external system. Give reference to the COTS product tables. If new tables are needed, give the table structure also.*>
6. **Process**<*Describe the process of receiving and validating the inputs, storing the processed inputs, receive parameters for generating the outputs, the process of converting inputs/stored information into outputs, protocols of transmitting the outputs to the output media in this section.Use diagrams where necessary. The diagrams may be flow charts, DFDs, Use Cases or any other applicable diagrams approved for use within the organization*>
<*Closing pages and appendixes, if any*>   <*This section may contain acknowledgements, list of executives that worked on the SRS, the list of documents used in deriving SRS, list of statutes referred to and so on.*>

5.7 Establishment of Requirements in Software Maintenance Projects

Software maintenance projects could be handled either by the in-house team or may be entrusted to an external organization specializing in software maintenance. When an in-house team handles software maintenance, the project would be handled using a maintenance work order (referred to as program change request, function modification request, software change request and so on in different organizations) which would contain details of the required software maintenance. If the project is entrusted to an external organization, then both the organizations would enter into a contract describing the process for resolution of software maintenance work orders, the turnaround times, prioritization policies, cost estimation, guidelines for code maintenance, and payment for the work carried out etc. Then each work request would be handled in conformance with the contract. Software maintenance work is classified under two heads, namely, software modification (including bug/defect fixing, modification to meet a changed requirement, or effecting some change to improve productivity etc.) and functional expansion. For software modification, the process would be as follows:

1. Receive the request
2. Replicate the described scenario in development environment
3. Locate the piece of code needing modification
4. Effect the required code modification conforming to software maintenance guidelines
5. Subject the modifications to quality control activities
6. Deliver the fix.

Some of the maintenance requests would be so urgent/small as to deliver in two hours. Others would have to be delivered within one business week at the most. In view of the short turnaround times, there is not much scope for formalized analysis and documentation in software modification type of maintenance work.

Functional expansion is normally adding additional functionality to the existing software and would involve building new code rather than code modification. This normally would be handled as a fresh software development project. However, such functional expansion projects would be of relatively shorter duration and would range from one calendar month to a maximum of one calendar year. The requirements establishment procedure and templates described in the foregoing sections for software development projects would be utilized for this kind of software maintenance scenario.

5.8 Establishment of Requirements in Migration, Porting and Conversion Projects

Migration projects involve migration of an existing system from one version of a software platform to a newer version of the same platform. The objective of these projects is to make use of the better facilities available in the new version which may include better response time, better user interface or better integration with other systems.

Porting projects move the existing system from one hardware platform to another hardware platform keeping the same software platform. The objective of these projects is to make use of a more robust hardware platform to cater to an increased load and to provide better response times to users.

Conversion projects involve modifying the existing system by keeping same hardware and software platforms to implement some major changes in the environment. Euro conversion and Y2K projects are excellent examples of this kind of projects. In the recent times, the introduction of Sarbanes–Oxley statute and the IFRS (International Financial Reporting System) have caused the existing systems to be converted to be compliant with the requirements of these statutes.

The execution of these projects has common software engineering cycle:

1. Develop/purchase a tool that migrates/ports/converts the existing code to the new platform. When a new software platform is released, it usually includes a migration tool for migrating code from older versions. Similarly for porting and conversion also, tool vendors quickly come up with tools. But it has been the experience that most tools do not complete migration/porting/conversion to the extent of one hundred percent. They do leave some gray areas untouched. These remaining pieces of code need to be migrated/ported/converted by hand.
2. Test the code/manually inspect the migrated/ported/converted code and modify any code that has not been fully migrated/ported/converted by the tool.
3. Subject the new code to applicable quality control activities.
4. Implement the new system in place of the older system.

Now in these projects, requirements engineering or management has a small role. The requirements are available publicly and tools are also mostly, available. The functionality of the product/system remains unaltered. Therefore, no templates or processes are described here for the establishment of requirements for these projects.

5.9 Establishment of Requirements in Agile Development Projects

Agile is a philosophy rather than a single software development life cycle. Many software development methodologies are shown under the umbrella of agile methodologies. Some of them are XP (Extreme Programming), Scrum, DSDM (Dynamic Systems Development Method), Clear Case, Feature-driven development, Test-driven development and so on. The philosophy is stated in the agile manifesto, thus:

1. Value individuals over processes and tools
2. Value Working software over comprehensive documentation
3. Value customer collaboration over contract negotiation
4. Value responding to change over following a plan

In line with the above philosophy, agile adherents do not encourage documentation and keep it to the minimum. However, "user story" is used as the main technique for recording requirements. True to agile philosophy again, no template is used for documenting the user story. As the phrase indicates, it is a story narrated by the user to describe the problem for the proposed product/system.

Therefore, it would be in the style of the user and the format and content differ from one user to another. It is possible that the user story may not provide all the information required for the developers. In such cases, the gaps in the information are obtained through interactions with the co-located customer. The agile methodology mandates "co-location" of the customer representative with the development team. So, the customer interactions are possible as and when required by the development team. Projects adhering to agile methodologies are able to deliver results to their customers proving that agile philosophy is indeed working. Therefore, I have not provided any template or format for establishing requirements in agile development methodologies.

Chapter 6
Quality Assurance in Requirements Management

6.1 Introduction to Quality Assurance

The word "quality" has multiple meanings and connotations. It means "character", "inherent feature", "degree of excellence" and "distinguishing attribute" among others. In the present context, we refer to "degree of excellence" as the meaning of the term "quality".

Quality is a term that is often used without any adjectives like good or bad. We often say "It is a 'quality' product" meaning that the product has superior or better than average quality.

International Organization for Standardization (ISO) defined the word "quality" as "the degree to which a set of inherent characteristics fulfills requirements. The term 'quality' can be used with such adjectives as poor, or good or excellent. 'Inherent' as opposed to 'assigned' means a permanent characteristic existing inside something."

Another popular definition of "quality" by Joseph Juran is "quality is fitness for use".

These two definitions have practical implications for implementation. Juran's definition leaves the terms "fitness" and "use" undefined. Unless these two terms are defined, the definition is open to misinterpretation by users and producers of goods and services and they often do. The ISO definition makes "quality" a continuum and usage of adjectives is mandatory. But most of us are used to using the term "quality" without any adjectives to mean good quality. Again the terms "good", or "bad" or "poor" or "excellent" are vague terms not lending themselves to quantification or measurement.

M. Chemuturi, *Requirements Engineering and Management for Software Development Projects*, DOI: 10.1007/978-1-4614-5377-2_6,
© Springer Science+Business Media New York 2013

Definitions cannot have vagueness that is open for misinterpretation. I offered the below definition for quality in the book "Software Quality Assurance: Best Practices, Tools and Techniques for Software Developers"[1] thus:

Quality is an attribute of a product or service that is provided to consumers, completely conforming to or exceeding the best of the available specifications for that product or service. It includes making those specifications available to the end user of the product or service.

The specifications that form the basis of the product or service being provided might have been defined by a governmental body, an industry association, or a standards body. Where such a definition is not available, the provider may define such specifications.

This is a more comprehensive definition of the term "quality". It mandates the provider to declare and conform to a set of applicable specifications. The specifications have to be defined by a national or international or industry association or a professional body. Only when such specifications are not available, the provider is free to define a new set of specifications. It mandates that the specifications cannot be secret or hidden from public view. It is also mandatory for the provider to meet or exceed those specifications. This definition of quality defines the minimum acceptable level for quality and that is to meet the specifications which are publicly available.

Having understood the term "quality", we can now move forward to quality assurance.

When manufacturing and services on commercial terms began, the quality depended on the capability of the artisan providing them. It was person dependent. But when production of large quantities of products began, an independent inspection was introduced to ensure that all parts were indeed included on the product. In due course, testing of the product was also introduced to ensure functioning of the product. These two activities, namely, inspection and testing, came to be referred together as "quality control" connoting the control of quality of outgoing products or services.

Even though quality control did a great job in ensuring that manufacturing activities build-in quality, the product failures still occurred due to poor specification or design. Quality control could not ensure that the pre-shop-floor activities, including product specifications and product design, are indeed building quality into the product. This led to introduction of standards for specifications and design. Verification and validation of specifications and design were started to ensure that quality is built into the product right at the stages of product specification and product design. Quality assurance includes all activities, that is, quality control activities and standards that ensure that quality is built into a product or service.

Now we have to acknowledge one fact that quality control activities, while they consume resources, do not help us build quality into the product. Therefore,

---

[1] Software Quality Assurance: Best Practices, Tools and Techniques for Software Developers, J.Ross Publishing, Inc, 2010.

the new concept of Total Quality Management (TQM) gained ground. TQM suggests building all activities necessary to build quality into the process (either for manufacturing or services) itself so that the need for quality control is drastically reduced and thereby reduces the cost of achieving quality itself. This has led to organizations moving towards a process-driven management.

This is a brief introduction to the vast subject of quality. Interested readers are suggested to read my book on software quality assurance cited in the foregoing section.

6.2 Quality Assurance in the Context of Requirements Engineering and Management

In order to ensure that quality is built into project requirements, we need to put in place five things:

1. Process, standards and guidelines
2. Right people with right training
3. Quality control of the deliverables of requirements engineering
4. Measurement and analysis
5. Project postmortem during the project closure.

Now let us discuss these five things in detail.

6.2.1 Process, Standards and Guidelines

These are in fact the pre-requisites for ensuring quality in the deliverables of the requirements engineering and management activity. As we have learnt in the earlier sections that the quality control activities are post-facto and do not build in quality into the deliverables. We have to ensure that the activities are carried out in such a manner that quality gets built into the deliverables by the individuals performing the activities. This is the philosophy advocated by TQM. This can be achieved by defining a comprehensive process for carrying out the activity in the organization and then internalize it among the individuals performing the requirements engineering activities. A process is a network of procedures, standards, guidelines, formats, templates and checklists which are appropriate for the organization. By appropriate, I mean:

1. They must be suitable for the kind of work being executed within the organization
2. They must be elaborate enough to ensure capture of complete information
3. They must be so designed as to assist the users in performing the work efficiently and comprehensively

4. They must be designed to be self-explanatory needing little or no training for use by trained individuals.

Procedures are step-by-step instructions and explain how to perform/accomplish an activity. Standards prescribe the parameters for the performance of the activity. A guideline is similar to a standard but is more suggestive and less prescriptive in nature than standards. Formats and templates assist the performers in comprehensively capturing the information. Format and template are similar in nature except that a template contains helpful hints to aid the users in capturing the information within the document. A checklist is tickler for the memory to alert the individual performing the activity to any missing information. The process ties up all these pieces together to present a comprehensive whole.

A comprehensive process for requirements engineering and management would include the following artifacts:

1. An overall process document linking all the artifacts that deal with requirements engineering and management with in the organizations
2. A set of procedures for:

    a. Elicitation of requirements
    b. Gathering of requirements
    c. Analysis of requirements
    d. Establishment of requirements
    e. Measurement and analysis for the performance of requirements engineering and management activity
    f. Verification and validation of the deliverables of the activity of requirements engineering and management

3. A set of formats and templates for

    a. Capturing the information during elicitation and gathering of requirements
    b. Document the URS
    c. Document the SRS
    d. Defect reporting during verification and validation activities

4. Checklists for:

    a. Elicitation and gathering of requirements
    b. Documenting URS
    c. Documenting the SRS
    d. Verification and validation.

Now let us not misunderstand that all the above must be paper-based documents. It could be soft copies or could be embedded inside a software tool.

6.2.2 Right People with Right Training

This is another pre-requisite to ensure that quality is in-built in the deliverables of the requirements engineering activity.

Who are the right people to carry out the requirements analysis? A few years ago, senior programmers, that is, the programmers that had put in a minimum of 2–4 years of experience in programming work were initiated into requirements analysis work under the close supervision of a project leader. Those were the days of mainframe computers. It continued until recent times. The advent of COTS products like ERP, SCM, CRM etc. caused this practice to change. In the projects implementing the COTS products, functional specialists with training on the specific product were utilized to analyze requirements. These individuals worked in the respective functional areas, be it material management, HR, or marketing; then received training in the usage of the COTS product and worked on implementing the product at a couple of sites in the minimum. Now, owing to a shortage of these functional specialists, the practice has moved forward to utilizing people with MBA (Master of Business Administration) or an equivalent educational qualification to carry out the requirements engineering function with some training on the product and requirements engineering. Universities started offering courses in business analysis to bridge the gap between the demand and supply of business analysts. The function of managing requirements rests with the software project manager in most of the organizations.

Now, as the present scenario exists, all persons enumerated below are utilized by organizations to carry out requirements engineering activity:

1. Senior software engineers who worked on similar functional domains
2. Functional Specialists with training on the respective product, especially in the projects implementing COTS products
3. People with MBA or equivalent qualifications and training in the requirements engineering
4. Project leaders and project managers

Who among these are the right individuals to carry out this activity? We cannot prescribe only one set of qualifications and experience to handle requirements engineering in all types of projects. My recommendations are based on the type of software project, which are as under:

1. Full life cycle software development projects—It is advantageous to utilize either functional specialists or project leaders or project managers who have handled projects in the same functional domain earlier.
2. COTS product implementation projects—In these projects, we can utilize functional specialists or people qualified in business administration or business analysis with training on the respective COTS product.
3. Conversion/migration/porting projects—in these projects, functional domain has little significance. Therefore, for these types of projects, senior software

engineers, and project leaders deliver the best results when the requirements engineering activity is entrusted to them.

4. Software maintenance projects—again, in these projects too, functional domain has little significance. Therefore, it is best that senior software engineers carry out the requirements engineering activity.

5. Testing projects—To test a software product effectively, knowledge of the functional domain is essential. Therefore, functional specialists or project leaders who handled projects in the same domain earlier are best suited for carrying out the requirements engineering activity.

Having right people is one thing but keeping them on the cutting edge of current technology is another. Their skills need to be continuously honed. The role of training in keeping the individuals up to date cannot be overemphasized. We normally conduct the following types of trainings to requirements engineering professionals:

1. **Induction Training**—We need to conduct induction training when a professional joins our organization. He/she might be proficient in the subject of requirements engineering but he/she needs to be trained on the organizational process, tools and techniques used in the organization. This will enable the person to deliver results that are consistent with the results delivered by the existing professionals.

2. **Tool Training**—we need to conduct training on the usage of tools and techniques used in the organization for carrying out requirements engineering activity. This training is conducted initially when an individual joins a new organization and when a new tool is acquired by the organization.

3. **Requirements Engineering Training**—Often, organizations promote existing programmers to take on higher responsibilities. Some of those promoted may be able to handle requirements engineering activity. In such cases, they need to be trained on the theory and practice of requirements engineering to prepare them for shouldering the responsibility. They may be trained in-house or sponsored to an external training program. If the individual is trained at an external institution, the person needs to be put through the induction training discussed above so that the organizational practices are imparted to the person.

4. **Continuing Education**—We conduct training periodically on the new developments in the field of requirements engineering. The periodicity of this type of training varies from organization to organization and on the developments taking place in the field.

5. **Knowledge sharing**—As we execute projects, the individuals gain fresh insights and knowledge about the discipline. We organize this type of training whenever a project is completed. This will be conducted by the persons who worked on the completed projects sharing the experience gained on the project. It would include sharing of the information on the successes and failures; best practices and pitfalls; any special developments of the project. This will enable all the professionals in the organization to have knowledge of all the projects executed in the organization even though they did not work on all the projects.

6. **Seminars, workshops and tutorials**—Professional bodies, universities and research institutions conduct seminars, workshops and tutorials on various topics of relevance inviting public participation. We nominate our professionals when these are on the topic of requirements engineering. From these, our professionals would gain insight into the experience of other organizations as well as the discoveries of the latest research. We cannot provide this knowledge through in-company training programs.

7. **Participation in discussion groups and message boards**—we encourage our professionals to participate in message boards and discussion groups on the topic of requirements engineering. These Internet forums provide for exchange of information freely among participants. In fact, these have become the primary source of knowledge improvement because the question can be specific and pointed to the issue at hand and relevant answers come forth. Such forums are available on Yahoo, Google, MSN, and LinkedIn among others.

8. **On-the-job Training**—Even when a professional is trained in the discipline, the person has a first project! During such occasions, the individual would be closely supervised by a senior specialist to provide the person with first project experience besides ensuring the success of the project. We used to refer to this as "apprenticeship" earlier.

We adopt all of these vehicles for training in organizations. In addition to these training programs, we also conduct project-specific training whenever a special situation exists in the project.

6.2.3 Quality Control

We carry out these activities to ensure that quality *was* indeed built into the deliverables. Verification and validation are the main activities for carrying out quality control activities. These are separately discussed in subsequent sections of this chapter.

6.2.4 Measurement and Analysis

We do not know the efficacy of the performance if we do not measure the performance and benchmark it with similar performance within the organization or outside the organization. This is a vital activity and a separate chapter is dedicated for a detailed discussion of this topic.

6.2.5 Project Postmortem

Normally organizations are focused on delivering results and completing projects rather than on analyzing the past performance. If a project is successfully completed, it is argued that, its performance is acceptable. But the real efficacy of project performance can be gauged realistically only when the project postmortem is diligently conducted. Project postmortem looks at the project performance from all angles in a critical manner to uncover best practices and pitfalls experienced during the project execution. This will add significant value to the organizational knowledge base and to the individual competence.

6.3 Verification

Verification is ensuring that the "right thing is done". It does not involve powering up/running the product and testing its functionality. Verification does not involve any tests to ensure that the right thing is accomplished. Verification is carried out visually and perhaps through touch and feel. IEEE standard 610 for "Standard Glossary for Software Engineering Terminology" defines the term "verification" as "the process of evaluating a system or component to determine whether the products of a given development phase satisfy the conditions imposed at the start of the phase" and also as "formal proof of program correctness". CMMI model document defines the term "verification" as "confirmation that work products properly reflect the requirements specified for them. In other words, verification ensures that 'you built it right'."

One thing is common in all these definitions—it does not involve ensuring the achievement of functionality by testing the product.

When we consider the case of requirements engineering, we do not have a product yet. So the question of testing the functionality does not arise. We have two types of verification that can be used in carrying out quality control of requirements. These are:

1. Peer review

   a. Independent review

      i. Individual review
      ii. Group review

   b. Guided review

      i. Individual review
      ii. Group review

2. Managerial review

    a. Independent review
    b. Guided review

6.3.1 Peer Reviews

Peer review is carried out by a person who has similar knowledge and experience as the author of the artifact. Peers are selected normally from within the organization but can be selected from outside the organization when it does not have the persons with the requisite experience and expertise besides the author. Peer review looks very closely at the contents of the artifact. Every aspect of the artifact receives attention including documentation guidelines, formatting, spelling and grammar in addition to the technical content. The objectives of peer review are:

1. Ensure that the technical content is accurate.
2. Ensure that the technical content is comprehensive.
3. Ensure that all organizational standards and guidelines are adhered to in the preparation of the artifact.
4. Ensure that the artifact is clear and lucid when used by others and that no ambiguity is present in the artifact.
5. Ensure that the content in no case contains unnecessary content whether it is innocent or malicious.

Figure 6.1 presents the peer review process pictorially.

Individual reviews are carried out by a single reviewer who is qualified and has the domain knowledge to review the artifact. In this case, the author and the reviewer interact with each other directly.

Group reviews involve multiple persons in the review. In group reviews normally a person acts as the review coordinator. Usually, the author or the project manager of the project acts as the review coordinator. The review coordinator selects the reviewers; provides them the artifact; collects feedback from al the reviewers; collates the feedback; arranges for implementation of the feedback and reviews the implementation of the feedback and passes the document to the next stage.

Independent review is carried out in the absence of the author of the artifact. The artifact is provided to the reviewer or reviewers. Reviewer/reviewers would review the document and provide the review feedback to the author/review coordinator for implementation in the artifact. Once the feedback is implemented in the artifact, the reviewer/review coordinator would verify the implementation and pass it for next stage.

Guided review is carried out in the presence of the author of the artifact. If only one reviewer is assigned to review the artifact, the author would present the details of the artifact to the reviewer. The reviewer provides the feedback to the author during the review itself. The author implements the feedback later on and presents

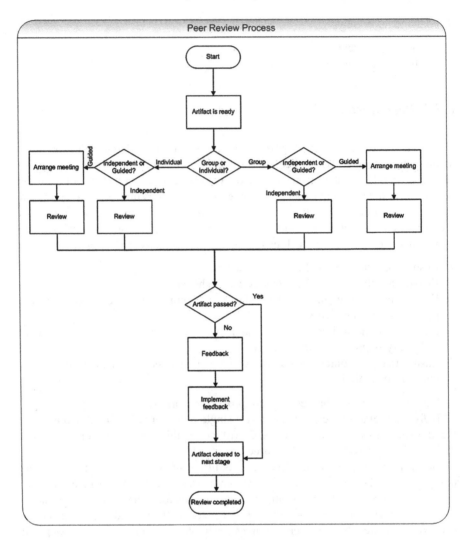

Fig. 6.1 Peer review process

it once again to the reviewer and obtains the go-ahead for the artifact for the next stage.

If multiple reviewers are selected for guided review, the review coordinator arranges a meeting of all the reviewers. The author presents the details of the artifact to the group. The review coordinator takes down all the feedback provided by the reviewers during the presentation and collates it. The feedback is received and implemented by the author of the artifact. The review coordinator reviews the implementation and passes it on to the next stage.

Table 6.1 Review feedback form

Review Feedback Report

1. Project name:
2. Name of the artifact being reviewed, with version number:
3. Name(s) of the reviewer(s):
4. Name of the author of the artifact:
5. Date(s) on which the review is conducted:
6. Type of review: independent / guided   individual / group   postal / meeting

Defects uncovered during the audit (use an additional sheet if required)

Defect ID	Defect description	Reference to process for the defect	Defect origin	Closed on	Status (open / closed)

Signature of the lead reviewer:
Date of signature:

Corrective actions implemented

Corrective action implemented	Defect IDs covered by this corrective action	Any comments

Signature of the author:
Date of signature:

Defect closure actions (to be filled in by the lead reviewer)

I have verified and found that all the defects described above are closed satisfactorily, except the following defects, which are retracted or pending

1.
2.
3.

Signature of the lead reviewer:
Date of signature:

Table 6.1 provides a suggested format for recording the feedback of the review. It is possible that the feedback is directly fed into a defect reporting tool, It is not necessary that the feedback is in paper form.

6.3.2 Managerial Reviews

Managerial reviews are carried out by the person to whom the author of the artifact reports. In other words, managerial review is carried out by the person approving the artifact. The objectives of carrying out a managerial review are:

1. The artifact submitted for approval is the right one belonging to the right project and product
2. The artifact submitted is complete in all respects and that no required information is missing
3. All quality control activities are performed; all feedback is implemented and that the artifact is passed for approval
4. Using the experience and the well-honed hunches, identify all possible problem areas in the artifact and correct them.

Managerial review is a bird's eye view of the document. It does not go into the minute detail as the peer review would. The person carrying out managerial review would glean through the artifact to see if everything is alright. Most often, managerial review would not uncover any defects. But if it discovers any defect, it would normally be a big issue necessitating a major revision. Manager, because of their experience gained in handling multiple projects in diverse domains are excellently positioned to uncover major slips of detail. But the major contribution of a managerial review is to ensure that all preceding activities including quality control activities are completed successfully.

Managerial review does not produce a review feedback form. The corrections are communicated to the author. The author normally subjects the artifact to peer review once again to ensure that it is reviewed in detail.

The deliverable for a managerial review is the approval of the artifact to the next stage, if no feedback is necessary.

6.3.3 Best Practices and Pitfalls in Verification

One pitfall is to treat verification as a mere formality to fill a review feedback form showing no defects for the purpose of quality audits and record keeping. We need to keep records for facing quality audits successfully but cooking up records is a bad practice. Verification adds value following the adage that "two heads are better than one". One is always blind to one's own faults/defects. So, a peer review overcomes that natural weakness of an individual.

The Second most common pitfall is to skip one of the two types of reviews. Both add values in their unique way. Often, peer review is skipped in preference for the managerial review. Managerial review in spite of the best intentions of the manager, would not be able to delve deep enough into the detail a peer review can.

Nor can peer review bring in the well-developed hunches and experience that a managerial review can. Therefore, neither review can really be skipped.

In some organizations, group reviews are omitted altogether. True, group reviews place an extra overhead on the project and the adage "too many cooks spoil the broth" may become applicable. Group reviews come in handy especially in projects handling new domains or of very large size. Group reviews have their place and should be utilized in applicable cases.

6.4 Validation

Validation is confirmation, authentication or corroboration of a claim. A claim may be that something is defect-free or something is working as intended. Validation confirms or rejects the claim.

IEEE standard 610 in its standard glossary of software engineering terminology defines the term *validation* as "The process of evaluating a system or component during or at the end of the development process to determine whether it satisfies specified requirements."

The CMMI® model document for development defines validation as "Confirmation that the product, as provided (or as it will be provided), will fulfill its intended use. In other words, validation ensures that 'you built the right thing.' The purpose of Validation is to demonstrate that a product or product component fulfills its intended use when placed in its intended environment."

Validation includes powering up the appliance or product and testing its functionality. Normally, validation is understood to be testing in the industry. But in requirements engineering, we do not yet have a testable product. Therefore, we need to adopt other means of validation. Here are some of the techniques that help us in validating the requirements:

1. Brainstorming
2. Story boarding
3. Prototyping
4. Expert review
5. End user review.

Let us discuss each of these techniques in greater detail.

6.4.1 Brainstorming

Brainstorming is a technique wherein concerned and knowledgeable people gather in an informal environment to give vent to free thinking. This is useful when a specialist expert is not available for the subject at hand. Everyone in the group shares their knowledge so that collectively the knowledge would be substantial.

For validating the requirements specification documents, the group deliberates the contents and comes to the best possible validation.

6.4.2 Storyboarding

Storyboarding is a technique used in film making in which pictures of the proposed movie are pinned on the wall sequentially so that the storyline is clear to the viewers of the pictures. If there are any gaps or absurdities present, they will come out clearly as well. This technique is used in validating the requirements documents. Requirements are printed on paper and are pinned in a logical sequence on a wall or a board. Then selected experts would go through the requirements and see if the requirements are valid requirements. This technique alleviates the necessity to read voluminous documents, make sense out of the write-up and decide the validity of the requirements. This technique is very effective but it would need extra effort in preparing the storyboard. Some people use a slideshow or PowerPoint slides to present the story instead of pinning the requirements on a wall or a board. The experts can view the slideshow any number of times to validate the requirements. This technique is used across many organizations for validating requirements.

6.4.3 Prototyping

Prototyping, as discussed in Chap. 5 on establishment of requirements, is used more to prove design than to establish software project requirements. But a few organizations still use, in special circumstances, prototypes to establish requirements and to validate requirements as well. Prototypes are especially useful in validating the product specifications (SRS). In product development projects, a prototype is built and it is demonstrated to select end users to collect their feedback. This feedback helps the organization to improve the product design as well. End users would not be able to visualize aesthetic aspects from requirements documents, especially of user interfaces of products. And as user interface plays a vital role in the success of the product, a prototype comes in handy to evaluate the user interface before we really built it.

6.4.4 Expert Review

Expert in this context is a person who is well versed in the functional domain of the proposed project or product. In some cases, the customer depends on the software development organization's expertise to deliver a software product with industry best practices included in it. In the case of product development, normally a functional expert guides the development team but to take advantage of another expert's knowledge as well as to validate the product specifications, another independent expert is requested to review the requirements and validate them. The expert may be a single individual or a set of individuals. The method may be an independent review or a guided review. Expert review uncovers gaps in the requirements, if any, and help in arriving at a comprehensive set of requirements.

6.4.5 End User Review

End users are the individuals who are likely to use the resultant software product from the proposed project. End users are found within the client organization for a client-oriented project or in another department of the same organization for an internal project. For business-to-customer projects such as airline ticket reservation, the end users are likely to be scattered across the world. In such cases, we need to carefully select end users from different demographic/geographical regions to get the best feedback. These end users ought to be knowledgeable about the document conventions and be able to understand the documents and provide intelligent feedback. The review may be independent or guided but in most cases, a guided review is preferred as the individuals are not part of the organization and hence are not obliged to abide by the timetable of the organization. Arranging a meeting and get them all together and present them with the requirements for review would be better as their feedback can be collated in one go. It is also effective on timeline considerations. A guided review also facilitates in ensuring that the end users understand the requirements as intended without bringing in their individual experiences to interpret the requirements.

6.4.6 Feedback Mechanism from Validation

The feedback vehicle presented in Table 6.1 can be used for collating and tracking the feedback to resolution. It can be paper based or software tool based. In the present day, the shift is more pronounced towards a software tool based feedback reporting and resolution mechanism.

6.5 Determination of Applicable Quality Control Activities

As can be inferred from the above discussions, there is some leeway and freedom regarding the choice of the quality control activities as well as how to perform them. The process at the organizational level sets general guidelines for project level quality control activities. At the project level the project manager prepares various plans including SPMP (Software Project Management Plan), SCMP (Software Configuration Management Plan), SQAP (Software Quality Assurance Plan), Project Induction Training Plan, Project Schedule and so on. Among these plans, SQAP would contain the quality control activities that are planned for the project. The activities needed for ensuring that quality is indeed built into the requirements documents are defined in this SQAP. Quality assurance activities to ensure that the project activities are performed so as to build quality into the requirements documents are defined either in the SPMP or SQAP deriving them from the organizational process, standards, guidelines, formats, templates and checklists for quality assurance.

Implementing the quality assurance as well as quality control activities is carried out during project execution including tracking and control using the usual project control mechanisms defined in the SPMP. These include status review meetings, status reports, measurement and metrics.

Chapter 7
Planning for Requirements Management

7.1 Introduction to Planning

The quote attributed to Abraham Lincoln states, "If I am given six hours to fell a tree, I will spend the first four on sharpening the axe", emphasizes the importance of planning. Another quote attributed to Peter Drucker states "Nobody plans to fail; but they just fail to plan". I don't think that I can do any better than these two quotes to explain the importance of planning. Planning may not prevent failure altogether but it certainly increases the chances of success and reduces the risk of failure in any human endeavor. It helps in anticipating the required resources reliably and in locating the risky areas in our endeavor so that we have adequate time to prepare and mitigate them successfully. But before we move further on this topic, let us define planning and understand its import.

7.2 Definition of Planning

Planning is defined as the intelligent estimate of resources required to accomplish a predefined endeavor successfully at a future date within a defined environment.
The key aspects of the above definition are:

1. **Estimate**—it is the best guess/anticipation of future requirements. It can be based on the organizational norms and the expertise/experience of the estimator.
2. **Resources**—cover the 4 M's, "Men (people, term 'men' is only used for the sake of rhyme), Materials including money, Methods, and Machines (equipment)". Resources are always applied over the course of the performance of the human endeavor.
3. **At a future date**—the dates for executing the project are in the future.

M. Chemuturi, *Requirements Engineering and Management for Software Development Projects*, DOI: 10.1007/978-1-4614-5377-2_7,
© Springer Science+Business Media New York 2013

4. **In a defined environment**—the environment where the work is going to be performed is defined. The environment is defined during the planning exercise. Any variation in the environment would affect the plan. The Environment refers to a wide variety of conditions including work logistics, workstation design, technical environment, processes and methods of management, prevailing morale at the workplace and corporate culture to name a few aspects.
5. **Endeavor**—any activity performed by human beings that would consume resources and would achieve a pre-defined objective.

The scope of the project would not be completely understood until we compile the requirements. But, what are the objectives of planning for requirements management? There could be multiple objectives for the endeavor. Here are the objectives of planning for requirements management:

1. Identify the resources needed to perform the requirements engineering activity efficiently and effectively. The resources are qualified and experienced human resources, money for expenses, duration needed and the machines (servers, laptops, tablets, PCs etc.) and tools required for the activity
2. Identify the risks associated with the activity and prepare a mitigation plan
3. Identify the dates for performing various requirements engineering activities during the course of project execution so that a project schedule can be prepared
4. Identify all the stakeholders of the activity so that every stakeholder contributes according to their role.

Any other project-specific objective can be added to the above list.

Having put the definition of planning in its proper perspective, we can now move forward on planning for requirements management.

7.3 Planning for Requirements Management

When do we carry out planning for a software project? This is normally the first activity as soon as the project is either approved or acquired. It normally precedes the requirements engineering activity. We prepare project plans during the project planning phase of the project. But in some cases, we may have to carry out some of the activities of requirements management even before planning the project. For example, in the case of internal projects, we may sometimes need to firm up the requirements as a prerequisite for the project to be sanctioned. A project would not be planned unless project is approved and budget is sanctioned. Even in external projects, we need to understand the requirements if only in their rudimentary state to offer a bid and acquire the project. Irrespective of where the project is executed, the organization conducts a feasibility study to ascertain if the project is worth committing funds to. The feasibility study would collect preliminary requirements to determine the need and the desirability of recommending the project.

What I have been trying to establish is that at least some portion of the requirements engineering activity precedes project planning, because no one would like to embark on an endeavor without any knowledge of what they are getting into. Such portion may be preliminary requirements as in most cases or as it is in a few cases, it could be full user requirements.

We can certainly plan for establishment of project requirements, change management of requirements, tracing and tracking of requirements through the course of project execution and measurement of the efficacy of requirements management in the project. We will discuss these aspects in the following sections.

7.4 To Document or Not to Document?

That is the question most people like an answer for! Documentation is not mandatory for planning. For small projects, documentation is not really essential. But for all other projects, documenting the plans offers several advantages:

1. It allows us to think through the project as well as review our own plans and ensure their comprehensiveness
2. It allows us to get a second opinion on the efficacy of our plans
3. In long drawn out projects, it allows us to keep the perspective. We are prone to forget the original plan in the long run; the document allows us to refresh our memory
4. It becomes the reference point to assess progress and control the project execution
5. We human beings do not have a photographic memory and tend to forget the details over a period of time. Plans impact multiple stakeholders and all would remember a different version of the plan in the long run, if it was not documented. If we wish to avoid arguments, differences of opinion and conflicts, we need to document the plans. A document acts as a point of reference and keeps every stakeholder on the same page at all times.

Therefore, it is advantageous to document the plans, especially in large and long duration projects.

Agile projects do not prepare planning documents because they follow an iterative development life cycle with each of the iterations being limited to a maximum of 4 weeks. In such situations, perhaps, a documented plan may not be necessary for each of the iterations. It is still advantageous to have an overall plan for all the iterations put together. Other projects do document project plans.

7.5 Different Project Plans

While full life cycle development projects prepare several plans, especially in large projects, we limit our discussion here to the plans relevant to requirements management. They are:

1. **SPMP**—Software Project Management Plan—this is the primary project plan and is referred to by many names including Software Project Plan, Software Development Plan, and Project Plan etc. This is the top level plan for a project and all other plans are subordinate to this plan. Other plans are spawned into separate documents only if the project is very large and deserves a separate plan for each of the other aspects of project management.
2. **SCMP**—Software Configuration Management Plan—this plan would include all activities focused on ensuring the integrity of all the code artifacts and information artifacts as well as controlling the changes to the artifacts that are subjected to the rigor of the organizational configuration management process.
3. **SQAP**—Software Quality Assurance Plan—this plan would include all activities focused on ensuring that quality is built into all the project deliverables. They include the quality control activities of verification and validation besides various processes, procedures, standards, guidelines, formats, templates and checklists selected for use in the project.
4. **Induction Training Plan**—this plan would include all activities focused on bringing newly allocated human resources up to speed quickly on the project. It includes the required topics of training for each role, duration for each topic, the methodology of imparting the training, faculty specifications, evaluation of the feedback received from the completed training programs and so on.
5. **Project Schedule**—this is a PERT/CPM (Program Evaluation and Review Technique/Critical path Method) based project schedule. It will include all the activities that need to be completed in order to execute and complete the proposed project. It will contain the sequence in which the activities need to be performed as well as the sets of activities that can be performed concurrently and those that need to be performed sequentially.

There are other plans like the implementation plan, master data creation plan, deployment plan, end user training plan, system changeover plan, software maintenance plan and so on that are used in especially large projects. But the five plans discussed above are most frequently used in the management of project requirements. Therefore, I have not delved deep into the details of the remaining plans.

Project planning is a large subject in itself and to cover it comprehensively is beyond the scope of this book. I have included sparse material about planning here as a prelude to planning activities relevant to the management of project requirements. Interested readers are advised to refer to "Mastering Software Project Management: Best Practices, Tools and Techniques" referred to earlier in this book.

7.6 Planning for Requirements Management in Projects

We need to plan for the following requirements engineering activities in our project plans:

1. Requirements Elicitation and Gathering
2. Requirements Analysis
3. Requirements Establishment
4. Requirements Change Control and Management
5. Requirements Tracing, Tracking and Reporting
6. Measurement and Metrics.

Let us discuss planning for each of these in greater detail.

7.6.1 Requirements Elicitation and Gathering

The activities of elicitation and gathering would be included in the SPMP which would include:

1. **The resources required**, namely, the human resources, the number required, the proposed role for each of the individuals, when exactly they would be needed (this aspect can be referenced from the project schedule), the likely release dates, the qualifications, experience and expertise necessary for the individuals, and any other project specific needs would be recorded here. Other resources like the laptops, voice recorders, video recorders, Internet connectivity, special software tools and any other project specific hardware resources would also be included in this plan. It would also include funds requirement, if any, for procurement of special hardware, software tools, travel, stay and boarding, as well as any other project specific requirements.
2. **The methodology** adopted for the project—The methodology includes the software development life cycle, work management (allocation and de-allocation, measurement and metrics etc.), communication, progress review, reporting and so on. SPMP would also record the specific methodology adopted for each of the aspects of the project. This could be simply a reference to relevant organizational process assets. Or it may point to documents maintained under the configuration management of the project. In some cases, it may even be recorded within SPMP itself! Whatever the case may be, SPMP will make it clear as to the specific methodology adopted for the project for elicitation and gathering.
3. **The standards and guidelines**—SPMP would normally specify the standards and guidelines selected for use in the project. These may be organizational standards or client organization's standards. SPMP would point to the location of the standards and guidelines selected for the project which may within the organizational process assets library or within the project configuration

management. Normally, standards and guidelines would not be recorded in the SPMP.

4. **The project specific processes**, procedures, formats, templates and checklists—SPMP would normally provide reference or pointers to formats, templates and checklists selected for use in the project. They would be either in the organizational process assets library or the project's configuration management.

5. **Project specific tools and techniques**—Projects would be using tools for preparing various diagrams like flow charts, DFDs, ERDs use case diagrams and so on. Other tools used in the project include configuration management tools, documentation tools (such as MS Office suite), data analysis tools, and communication tools including audio/video conferencing and collaboration tools. All these would be enumerated in the SPMP and pointers would be included to locate their user manuals or help files.

Thus most of the activities of elicitation and gathering are covered by the SPMP and the project schedule. The remaining activities would be covered in the other plans.

7.6.2 Requirements Analysis

This aspect would find a place in two plans, namely the SPMP and the project schedule. The project schedule would record the dates on which this activity would be performed. SPMP would contain the details of the persons carrying out the activity, the methodology, and the tools and techniques utilized in the analysis. SPMP would also contain the pointers to the procedure, standards, guidelines, formats and templates necessary for carrying out this activity.

7.6.3 Requirements Establishment

The activities that need planning are the resources, the timelines, the methodology, standards, guidelines, formats, templates and checklists. The methodology includes documenting guidelines, quality control methodology, approval methodology, and configuration management methodology. We also need to plan, the tools and techniques used for documenting, defect reporting, resolution, and communication.

SPMP would record the methodology and pointers to the documentation guidelines, adopted for the project in the establishment of project requirements. It will also include information about the selected tools and techniques proposed for use on the establishment of the project requirements.

SCMP would record the selected configuration management methodology for establishment of project requirements as well as the change management methodology. It will also record the version control methodology of project's software artifacts including check-in and check-out procedures.

SQAP would enumerate the quality control activities selected for the establishment of project requirements. It will include pass/fail criteria, defect reporting and resolution methodology as well as escalation methodology and mechanisms in case of disputed decisions.

The project schedule would contain various timelines for the establishment of project requirements, performance of quality control activities, list of artifacts needing approval from customer, and submission for and receipt of approvals.

7.6.4 Requirements Change Control and Management

Greek Philosopher Heraclitus said, "There is nothing permanent except Change." Hardly any project is ever completed without some changes being requested and implemented. So, it is very important that we plan for receiving and implementing change requests. Change management forms part of the configuration management process. Therefore, it is planned for in the SCMP. As it is rather not possible to predict the timelines for receiving the change requests, so, the project schedule would not contain any change management activities. The SCMP would however, contain all the activities pertaining to change management including receipt of change requests, recording them, analyzing them, implementation strategy, implementation, quality control activities, and measurement and metrics to analyze the impact of change requests on the project.

7.6.5 Requirements Tracing, Tracking and Reporting

Requirements are to be traced through all the activities of software development so as not miss any of them at any stage. This will ensure a software product that truly meets all the requirements of the end user as well as those of other stakeholders. Tracking and reporting of the requirements engineering activities provide needed information to all the stakeholders and keep them on the same page. Therefore, all these activities need to be planned. SPMP contains the information on the methodology, periodicity and tools used in tracing of the requirements through the software development. SPMP would also include information on the methodology used for tracking and reporting on the requirements engineering activities as well as the formats and templates used for reporting the performance of the requirements engineering activities in the project. SPMP would also enumerate the agencies responsible for tracing, tracking and reporting activities, the periodicity of reporting as well as the agencies that receive the reports. The storage of the documents used in tracing, tracking and reporting is defined in the SCMP.

7.6.6 Measurement and Metrics

Measurement facilitates determining the progress quantitatively, which facilitates drawing inferences and initiating necessary corrective actions commensurate with the variance. Metrics facilitates benchmarking the performance with other projects within the organization or with the other similar organizations. The aspect of measurements and metrics relevant to requirements engineering and management are covered in a separate chapter. SQAP normally contains all of the proposed measurements for the project and so would contain the measurements and metrics proposed to evaluate the requirements engineering activities. In some cases, SPMP would contain this information. The plan would contain various measurements, their periodicity, proposed metrics, the methodology to derive the proposed metrics and the formulas thereof, as well as the proposed analysis and benchmarking. It would also contain the information of how and to whom the metrics information would be communicated.

7.6.7 Formats and Templates for Planning

I have not provided the formats and templates for the plans discussed in this chapter as the information pertaining to requirements engineering forms a miniscule portion of these plans. The plans contain many more aspects of project planning and execution. Also, we do not make separate plans for requirements engineering activities and I do not advocate preparing separate plans for requirements engineering activities.

7.7 Best Practices and Pitfalls in Planning

I have seen planning being treated as an exercise in creating planning documents. This is the worst pitfall many organizations fall into. Planning is an exercise in thinking trough the projects and evaluating every task in terms of the resources it needs, the quantity, type and timelines the resources are required, the methodology that is best suited to accomplish it successfully, the timelines for its performance, and the tools and techniques necessary for its successful accomplishment. Documentation is to retain that thinking for reference throughout the duration of project execution by all the stakeholders. But planning is certainly not an exercise in creating documents.

Another pitfall that organizations frequently fall into is avoiding the documentation altogether. There are many stakeholders to a project that include end users, customers, organizational management, the project team, the quality assurance department, finance and marketing departments, the program manager and so on.

If all these agencies are to be brought on to the same page, nothing is more simpler and cost effective than a document. A format helps documenting the plan comprehensively than placing an overhead on the project manager. It reminds the project manager of any aspects of the project that are either missed or forgotten.

Another pitfall that I had occasion to witness is to treat project schedule prepared using MS-Project or such other tool as the entire plan for the project. Nothing can be more misleading. It is simply a schedule. Such schedules do not contain the selected methodology, standards, guidelines, formats, templates and checklist for the project and tools and techniques proposed for use on the project. A schedule also does not contain the methodology of carrying out quality control activities, defect reporting and resolution methodology, an escalation mechanism, nor does it document the configuration and change management procedures. So, preparing a schedule and treating it as the comprehensive project plan is not very wise.

A less common pitfall is to overdo the documentation part. Use of a very comprehensive template even for a short duration project is a pitfall. The plan ought to be commensurate with the size and duration of the project. Using the same comprehensive template for all projects without considering the project size and duration is not very wise.

Chapter 8
Requirements Change Management

8.1 Introduction

The quote, "There is nothing permanent except Change" attributed to Greek Philosopher Heraclitus, emphasizes how the world changes. World changes and requirements too, change midway through the project execution. Almost all projects would need some change or the other after the requirements are frozen.

Requirements change management begins only upon freezing of the requirements, that is, the requirements documents are approved and are subjected to the project's configuration management. It will continue through the project execution until the project deliverables are handed over to the customer.

What is a *change*? A change is basically a requirement that is specified/modified after the requirements are frozen. The new requirement may be a modified version of an already specified requirement. Or it could be a new requirement altogether.

Why do requirements change?

Requirements change midway through the course of project execution for a variety of reasons. Core functionality requirements may change due to:

1. The business environment in which the organization operates undergoes a change to which the organization needs to respond. This can cause the software requirements to change.
2. The management of the organization may effect a reorganization of its operations and this may necessitate a change in the core functionality requirements.
3. Some of the end users may have forgotten some requirements or remembered a new requirement after freezing the requirements. This would necessitate a change of requirements.
4. A new statute or a court judgment, or a government diktat can cause changes after the requirements are frozen.
5. Process improvement activities may have modified some of the existing business practices and these can cause the requirements to change.

M. Chemuturi, *Requirements Engineering and Management for Software Development Projects*, DOI: 10.1007/978-1-4614-5377-2_8,
© Springer Science+Business Media New York 2013

6. Data analysis of completed projects can reveal some anomalies which could result in changing the requirements.

Ancillary functionality requirements also can change for the following reasons:

1. During design phase or construction phase, it may be uncovered that implementation of some requirements as frozen may not be possible due either to technical reasons or cost considerations. This may necessitate a change of requirements.
2. Some of the system software may release patches or service packs affecting the software design causing the requirements to change.
3. A new threat or a security hole may be discovered in the system software necessitating a revision of the requirements.
4. Someone may uncover a better way to achieve the functionality causing the requirements to change.

Whatever the reason may be and whatever the requirement may be, some changes become necessary during the project execution after the requirements are frozen. Therefore, we need to equip ourselves with the means to handle changes in an orderly manner ensuring that the smooth work flow continues without major interruption. Normally this is covered in the SCMP and forms part of configuration management of the project.

8.2 Communication of Changes

How are changes communicated to the project team? Changes can be communicated to the project team either by telephonic information, in person, through an email, through a software tool or more formal methods. Agile methods mandate colocation of the customer with the project team and hence see no need for formal or written methods for communicating change. In the agile projects the communication would be in person. In other projects, all the above methods would be used. It is possible to communicate changes without written documents but, formal methods have advantages. They are:

1. Formal methods help in keeping a record of the changes requested for analysis later at the end of the project.
2. Formal methods facilitate tracking each change to its resolution and that no requested change is forgotten.
3. Formal methods ensure that all required information is communicated along with the requested change.
4. Formal methods enable analysis of changes and effect improvements in the process to minimize changes as well as to improve the process of defining project requirements more comprehensively.

The formal mechanism used for handling change management is a CR (Change Request). A suggested format for a CR is shown in Table 8.1.

8.3 Origination of Changes

Changes can originate from various stakeholders including:

1. **Customers**—Customers' representatives raise change requests mainly to change core functionality requirements. Occasionally they can also raise requests to modify ancillary functionality when the system software proposed by them has undergone changes. The changes stem mainly from changes in a business scenario, or new/modified statutes, and reorganization of key departments etc. The world is dynamic and anything could change to effect the frozen core functionality requirements and customers could raise change requests.
2. **End users**—End users raise change requests when a frozen requirement needs to be changed because they either forgot a key aspect of a requirement or they forgot a requirement totally. They may add a field or modify a field; they may change the screen layout or report layout; they may need an additional report; and modify the steps of a process and so on. Normally change requests raised by end users would affect core functionality requirements.
3. **Project team**—Project team members can raise a change request occasionally when they are not able to implement a requirement in its entirety or need a design modification. They may not be able to pack all the controls on the same screen or all the fields on the same report and this would cause them to raise a change request. Sometimes, they may be able to combine multiple screens into one screen layout. Normally project team's change requests are concerned with implementing the design and issues thereof.
4. **Testing team**—It is rare that a testing team raises a change request as it is focused more on uncovering defects than on finding opportunities for improvement. But testing teams may find opportunities for improvement (especially about system response times uncovered on performance testing, system stress uncovered in stress testing or concurrency control aspects uncovered in concurrent testing) while carrying out testing and may raise change requests albeit in practical terms these changes often are initially confused with problem reports. Testing teams do find some opportunities for improvement and raise change requests to resolve those changes.
5. **Organizational Standards group**—Organizational Standards groups may change an existing standard or bring out a new one, which may impact projects in progress. In such cases they may raise a change request to retrofit the standard into the project deliverables. Unless the change addresses a critical issue, the Organizational Standards group generally identifies a migration path for the change.

Table 8.1 Change request form

Requirements Change Request Form

Project Id:
Project Name :
Date :
CR Reference :

Initiator Information

Name of Initiator	
Designation	
Contact Information : Phone Number Email id Location	

Details of the Change Requested

Name of Module affected by the CR	
List of components affected by the CR	
Description of the requested Change (Add additional sheets, if required)	
Reasons for the change	
Priority (Immediate implementation / when possible before completion of project / to be retrofitted at the end of project)	

Implementation Information

Aspect	Name Of the Person	Date of completion
Analysis		
Approved for implementation or Rejected		
Implementation		
Review		
Regression Testing		
Closed on		

These are the sources for origination of change. Now let us look at the resolution of change requests.

8.4 Change Request Resolution

The first response to a CR is to record it in a Change Request Register (CRR) so that it can be tracked to resolution. A CRR could be as simple as an Excel Worksheet or a software tool like PMPal that facilitates the functionality of a CRR. The CRR is the main tool for tracking all CRs to resolution. By recording the CR in the CRR will ensure that no CR is overlooked/forgotten. It would also enable tracking every CR to resolution. It would further enable us to analyze the CRs at the end of the project. The resolution of a change request can be:

1. Accept the CR and implement it immediately
2. Accept the CR but implement it later along with all other CRs
3. Reject the CR

After a change request has been logged in the register, the CR is analyzed by a PM or a designated person. In large projects, there would be a CCB (Change/Configuration Control Board) that would analyze the CR and accord approval for implementation or reject it. In either case, the analysis would determine:

1. Whether the information contained in the CR is comprehensive with all pertinent details and facilitates implementation of CR.
2. Whether implementation of CR would be feasible both in technical as well as financial considerations. When the CR is raised by internal sources such as the project team or testing team, the analysis would also determine if the implementation is desirable from a user viewpoint in addition to its feasibility.
3. The amount of effort, cost and calendar time it would consume to implement the CR.
4. The impact of the CR on the overall project, if it is accepted (especially in terms of effort, schedule and cost) or rejected (fulfillment of functionality).

Once the analysis is completed, the Impact Analysis would be submitted to CCB or the PM who would approve or reject the CR. In the case of rejection, the decision along with reasons for rejection would be communicated to the originator of the CR and the CR is closed in the CRR. If the CR is accepted, the PM or the CCB would decide on the strategy for implementation of the CR. Once a CR is approved for implementation, it would be implemented in accordance with the CR implementation strategy decided and recorded in the Software Configuration Management Plan (SCMP). Strategies can include:

1. Implementing CRs immediately on receipt and approval
2. Consolidate all CRs and retrofit them at the end of the project or any another appropriate point in the development process

3. Situational implementation:

   a. If the component which is affected by the CR is yet to be constructed or is being constructed, then implement the CR when the impacted component is under construction.
   b. If the construction of the component, which is affected by the CR, is completed, keep the CR pending and retrofit it into the component at the end or at a convenient time.
   c. If the construction of the component is completed but not implementing the current CR would render the component a bottleneck for other components, it would be implemented immediately.

Once the analysis and strategy for CR implementation are decided, the CR would be implemented in line with the analysis and implementation strategy decided by the CCB or the PM.

Changes can cause disruption to the flow of project execution regardless of whether it begins as a smooth flow or a chaotic path. When a CR is received, in many cases, it would impact an artifact that is already completed. This causes severe impact. In rare circumstances the impact would be on an artifact that is yet to be completed. If the impact is only to the current artifact, the impact will tend to be less severe. Regardless of the scenario, artifacts like requirement documents, design documents and others will have to be reviewed and may have to be revised which will impact the project execution flow.

The phase of development at which the CR is received also determines the severity of the impact. For example, a CR is received just after the requirements phase will tend to cause the least severe impact as compared to a CR for the same item received when the project is in the system testing phase.

Table 8.2 summarizes the severity of the impact of CRs on project execution flow. But one thing is certain; CR does impacts project execution flow.

The timing when a change request is received can influence the implementation of CRs as much as the strategy. Table 8.3 shows the impact on a set of typical artifacts based on the phase during which the CR is received and the possible strategies for implementing the CR.

8.5 CR Implementation

First let us look at scenario where a project is using the most used situational approach to implement accepted CRs. The following are the steps generally followed for implementing the CRs under this strategy:

1. Is the activity impacted by the CR completed? If the activity is not completed and is yet to be started, the CR would be incorporated into the requirements and design (as required) documents.

Table 8.2 Severity of the impact caused by CRs

Phase in which CR is received	Severity of impact caused by CR based on the type of CR			
	URS CR	SRS CR	Design CR	Construction CR
User requirements	Nil	Nil	Nil	Nil
SRS	Medium severity	Nil	Nil	Nil
Design	High severity	Medium severity	Nil	Nil
Construction	High severity	High severity	High severity	1. Medium if the component is already constructed 2. Low, if the component is not yet constructed

Table 8.3 Impacted artifacts and the strategy for implementation of CR based on the phase during which the CR is received

Phase in which CR is received	Artifacts impacted	Suitable strategies for implementation
User/software requirements phase	User/software requirements documents	As and when received or when convenient but before design is started
Design phase	User/software requirements documents and design documents	As an when received or when convenient but before design is completed
Construction phase	User/software requirements documents, design documents and source code	As and when received or, retrofitted or situational implementation

2. If the work on the impacted component is started but not completed, the CR would be handed over to the team member (or members) carrying out the work for implementation. The CR would then be incorporated into the deliverables.
3. If the work on the component impacted by the CR is completed, then the CR would be kept pending to be implemented either at the end of the project or at a convenient time, such as when some resources become free or a part of the team is idle waiting for some approval or clarification and so on.

If the project was following a strategy of holding CRs and then retrofitting them at a convenient point in time, the following steps would followed to implement the CRs:

1. Each CR is further analyzed to determine the components and deliverables impacted by it.
2. At the completion of analysis, CR implementation activities are consolidated into packages normally by the component.
3. Work allocation would be made so that all CRs pertaining to one component or one set of related components would be allocated to the same set of team members.

4. The allocated team members would carry out the activities required to implement the CR.
5. The additions/modifications would be subjected to planned quality control activities such as reviews and regression testing.
6. All defects uncovered during reviews and testing would be rectified by the concerned team members.
7. Once all CRs are implemented, a managerial review of CR implementation would be carried out by the PM or a person designated by the PM to ensure that all the accepted CRs are satisfactorily resolved and they passed through quality control activities. Then all the CRs would be closed.
8. The artifacts would then be promoted to the next stage.

If the CRs were implemented when received or when convenient, the following steps would be taken to resolve the accepted CRs:

1. The CR would be allocated for resolution to the appropriate set of team members
2. If the CR impacts an information artifact,

   a. The information artifact would be copied to team's folders for modification so that the original will be unaffected.
   b. It would be modified as necessary.
   c. It would be subjected to quality control activities, namely the peer review and managerial review.
   d. Any defects uncovered during quality control activities would be rectified by the concerned team member.
   e. After all defects are rectified, the artifact would receive appropriate approvals.
   f. The current artifact in the configuration management folders would be moved to the archived artifacts folder and the updated artifact would be moved to the configuration management's current folders.
   g. All concerned team members would be informed of the change in the artifact.
   h. If the CR implementation includes modifying the code artifacts in addition to information artifacts, the CR would then be passed on to the team members assigned with the work of implementing the CR in the code artifacts along with reference to the updated information artifact.

3. If the CR impacts a code artifact, either independently or after an information artifact has been updated, the following steps would implement the CR in code artifacts:

   (a) The PM would allocate the CR for resolution to an appropriate set of team members for implementation along with references to any updated information artifacts.
   (b) The allocated team members would carry out the necessary coding. This activity would be governed by the coding guidelines for the project.

(c) The CR would then be allocated for Peer Review. The personnel involved in the Peer Review would review the code to ensure that the:

    a. Implementation fulfills the requirements of the CR.
    b. The implementation conforms to the project guidelines and other software engineering standards of the organization.
    c. There is no trash or malicious code left in the software.
    d. The changed code ensures efficiency of execution and response times.

(d) Once the CR is passed thru the Peer Review, it would be submitted for Regression Testing.
(e) The testing team would carry out regression testing to ensure that all changes and additions requested in the CR are correctly working and that the original functionality is unaffected by the implementation.
(f) Once regression testing is completed and all defects pointed out either in peer review or regression testing are resolved and closed, then the artifact is promoted to the next stage and the CR is closed in the CRR.

8.6 CRR

The CRR is used to record all the CRs received from any source and track each one to closure. The actual format of CRR can be an excel sheet or a tool based register. It is normally maintained electronically as soft copy.

The CRR would normally contain the following entries:

1. CR Reference number
2. Date on which the CR is received
3. Approval information including who approved it and the date of approval
4. Allocation details for Analysis including to whom it is allocated and completion date
5. Allocation details for implementation of CR including to whom it is allocated, and completion date
6. Allocation details for peer review including to whom it is allocated, and completion date
7. Allocation details for regression testing including to whom it is allocated, and completion date
8. Status—open, closed or under analysis/approval/implementation/peer review/ regression testing
9. Date on which CR is closed

Table 8.4 shows a suggested CRR format.

Table 8.4 Change request form

CR Ref.	CR Date	Approved By	Approval Date	Analyzed by	Analysis Start Date	Analysis End Date	Implemented By	Implementation Start Date	Implementation End Date	Peer Reviewed By	Peer Review Start Date	Peer Review End Date	Tested By	Testing Start Date	Testing End Date	Status	Closed Date

8.7 Progress/Status Reporting of CRs

Normally the status of implementation, progress of CR resolution and CR metrics are reported as a component of the Weekly Status Reports to the concerned executives. This serves the purpose of providing historical records and alerting senior management to the need for intervention where necessary or warranted.

8.8 Handling the Impact of CRs

Once we determine the impact of the CRs, we need to make a decision on how to handle the impact. We have the three options:

1. Absorb the impact. Ensure that the schedule and cost are not affected.
2. Pass on the impact to the customer completely. Customer will approve delayed deliveries, and pay extra for the additional effort spent on implementing the CRs.
3. Absorb partially for small CRs and pass on the impact of the bigger CRs to the customer.

One important aspects or one of the most important aspects is that we need to set a policy for handling the impact of the CRs and make it a part of the contract to ensure that the CRs do not cause a rift between the project team and the customer. More often than not, contracts miss this aspect of handling the impact caused by the CRs. Some of such organizations have ended up fighting in a court of law. I recommend the following strategy:

1. Absorb the impact of small CRs, that is, any CR requiring 2 person hours or less. But this is subject to a limit of 1 % of the estimated effort for the project. That is small CRs would continue to be absorbed until the cumulative effort spent in implementing small CRs reaches the absorption limit

2. Any CR needing more than 2 person hours, the impact would be passed on to the customer
3. The impact analysis in terms of effort, schedule and cost would be submitted to the customer for approval before implementing the CR. All CRs would be implemented only after approval by the customer.

Now if the project is internal, that is, the development team and the end user team work in the same organization, there would be no cost implication but the schedule implication would still be applicable. Since it is internal project, we should strive to absorb the schedule impact because the delay in deliveries would adversely affect the system roll out.

Now the above strategies would work fine for CRs from end users/customers. How do we handle the CRs from project team itself? Well, the change is ours and therefore, we cannot pass it on to anyone. We need to absorb it.

8.9 Measurement and Metrics of Change Management

Change Management is a critical activity of requirements management. I am dealing this topic in a separate chapter on measurement and metrics of requirements management itself.

Chapter 9
Requirements Tracing, Tracking and Reporting

9.1 Introduction

The most common cause of software product failure or dissatisfaction of end users with the delivered product is missing requirements or badly implemented requirements. Some of the requirements are forgotten, or changed, or deleted or poorly implemented especially in made-to-order software development projects. So, we need to trace, track and report the transition of requirements into a software product all through the software development life cycle. This chapter discusses these topics.

9.2 Requirements Traceability

Requirements tracing involves identifying the requirement in all the software artifacts including information artifacts and code artifacts.

Wikipedia defines traceability as "the ability to chronologically interrelate the uniquely identifiable entities in a way that matters".

IEEE standard 610—IEEE Standard Glossary of Software Engineering Terminology defines traceability thus: "The degree to which a relationship can be established between two and more products of the development process, especially products having a predecessor-successor or master-subordinate relationship to one another; for example, the degree to which the requirements and design of a given software component match" and also as "The degree to which each element in a software development product establishes its reason for existing; for example, the degree to which each element in a bubble chart references the requirement that it satisfies."

CMMI model document for development version 1.3 defines traceability as "A discernible association between requirements and related requirements,

M. Chemuturi, *Requirements Engineering and Management for Software Development Projects*, DOI: 10.1007/978-1-4614-5377-2_9,
© Springer Science+Business Media New York 2013

implementations, and verifications." and bi-directional traceability as "An association among two or more logical entities that is discernible in either direction
(i.e., to and from an entity)"

There could be other definitions of the term traceability offered by other
authors. The implications of the above definitions are:

1. We need to be able to trace the path of a requirement from its origin to the
   features of the end product.
2. We need to be able to trace the requirement in the reverse, that is, we should be
   able to trace a product feature to its origin in the requirements.
3. We need to be able to trace the requirement to any intermediate artifact in both
   directions that is from the requirement to the feature of the artifact and from the
   feature in the artifact to its origin in the requirements.

In short, traceability involves the ability to trace requirement to any intermediate point in the evolutionary path traversed by the product and from any point in
that path to its requirement.

9.3 Need for Requirements Traceability

Requirements can be changed at any point during the software development life
cycle as depicted in Fig 9.1.

Any stakeholder can place a change request at any point during the software
development life cycle. Some of the requirements may be deleted altogether. Some
fresh requirements may be added. The requirements at the stage of their definition
and the product testing stage may not be the same. If we wait till the end and try to
match the product features with the original requirements, it is possible that there
would be many mismatches. If we need to be able to prove that all customer
requirements are diligently met, we need to be able to trace them through the
software development life cycle.

Another aspect is the scope management for the project. The scope defined at
the beginning of the project would undergo changes due to the change requests
raised by various stakeholders. The scope in most cases would be increased and in
a few cases may even be reduced. But in most cases, the scope would not remain
the same. To keep track of the scope creep, tracing the requirements through the
software development life cycle would be very useful. We would be able to
identify when and how the scope of work has changed.

It has been generally accepted by the software development community, that
100 % testing is impractical even if it is desirable. In the scenario of changing
requirements, it would be difficult to assess the test coverage of the defined
requirements. If we trace the transformation of requirements through the software
development life cycle, we would be in a better position to assess the coverage of
requirements in our testing.

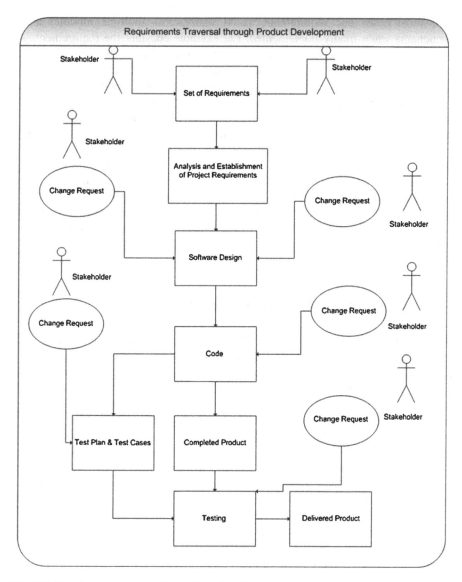

Fig. 9.1 Requirements traversal through product development

It is not uncommon for the project manager to change midway through the project execution due to a variety of reasons including resignation, termination, illness, allocation to another project and so on. When the new project manager takes over the project, it would be very difficult to understand where a feature's requirement originated without a proper traceability of the requirements. Diligent tracing of the requirement through software development life cycle would be a great help in such cases when the project manager changes.

In some cases one requirement would span across multiple modules and components when finally implemented. For example consider the requirement—"the system must be secure". It needs to be implemented in multiple components such as the login screen; prevention of accessing any screen directly bypassing the login screen; audit trails to investigate any intrusions; firewalls, anti-viruses and so on. Similarly, one component could be implementing multiple requirements. For example a sales transaction would trigger multiple actions across the system; the stock needs to be decreased; the income needs to be increased; the salesperson performance needs to be incremented; the customer may need to be created; the customer account needs to be suitably updated and so on. Unless we trace all requirements to implementation and all implementation to their requirements, we would not know how comprehensively did we implement the requirements.

Therefore, it is important and essential to trace all requirements in both directions, namely from requirement to implemented features and from product features to their requirements.

9.4 Mechanisms for Tracing Requirements

The most popular mechanism used for tracing requirements through the software development life cycle is the traceability matrix. An example of a traceability matrix is depicted in Table 9.1. In the traceability matrix, each requirement is traced in a row. The first column would contain the requirement id which is normally the one taken from the URS. The brief description column would contain a brief description normally taken from the requirements analysis sheet for consistency. References columns would be filled with the section numbers from information artifacts and program names from code artifacts. Each column may contain multiple references. In cases where a requirement is completely deleted, we indicate "Deleted" and include its reference in the column captioned "Deleted/Modified" column. If it is modified, the change request id is indicated in the "Change Request" column. This column too may contain multiple references. The description in the "Brief Description" column would not change when the requirement is deleted. When any requirement is modified, we need to evaluate if the description needs change and affect it on a case-by-case basis.

We may add more columns or rows depending on our own unique requirements. The traceability matrix is normally implemented using Excel worksheets. But if the number of requirements is very large, Excel sheets would be cumbersome to manage. We may need to use a specialized requirements traceability tool.

With Excel sheets, we can trace from either requirement to its implementation or from an implementation to a requirement. We can also use a separate worksheet for each module to cover the entire project in the same file. Excel is an excellent tool for implementing requirements traceability for a project. But when we come to the organizational level, Excel sheets would not be able to consolidate the information. It is better to use a specialized tool with RDBMS in the backend.

Table 9.1 Requirements traceability matrix

Requirements traceability matrix

Project id:

Last updated on:

Requirement id	Brief description	Origin of requirement	References of the requirement in other artifacts						Deleted/modified (include reference)	Remarks
			SRS	Design document	Code component	Test cases	Test logs	Change request		

9.5 When Should We Trace the Requirements?

We need to trace the requirements throughout the software development life cycle all the time. We need to be especially careful whenever a state transition occurs in the software development life cycle. Here are the typical stages where we need to update the traceability matrix:

1. Whenever a change request is implemented—usually most change requests impact requirements. So, every time we implement a change request, we need to see if any requirement has been affected and then update the traceability matrix with the resulting changes.
2. When we complete the software design for any module—first, we need to fill in the references of the design document against the implemented requirements in the traceability matrix. It is possible that some of the requirements might have been changed to suit the design. So, we need to assess if any requirement is impacted by the design and then update the traceability matrix.
3. Whenever a code component is completed—the coding of a component is the final stage of implementation as testing is merely confirmation that the implementation is defect free. It is possible that a component could have implemented multiple requirements either fully or partially. For example, the requirement of security is implemented across multiple components. So, as soon as we complete the coding of a component, we need to update the traceability matrix against all the requirements which were implemented in the component.
4. Whenever we complete unit/integration/integration testing of a component/ module/product—usually, testing does not impact a requirement. It merely confirms that the implementation is as defect free as possible. But there are occasions when the testing uncovers a major defect due to which we may have to re-visit not only the code or design but also the requirements. After a test is completed, we need to assess if such a situation has arisen due to testing and update the traceability matrix suitably.

As we update the traceability matrix, it would become easier for us to trace any requirement in either forward pass (passing from requirement to implementation) or backward pass (implementation to requirement).

9.6 Tracking of Requirements

What is the big difference between tracing and tracking of requirements? Is it different at all?

"Yep" I would say. Tracing is just ensuring that every requirement is implemented in all stages of software development. Tracing requirements through the software development life cycle ensures that the requirement has traversed all the stations in its path to implementation. But, what about the qualitative aspect of implementing the requirement, does tracing ensure it? How well is the requirement implemented?

Tracking requirements ensures that the requirement is implemented comprehensively both in letter and in spirit.

Tracking requirements requires diligence from project management team. It is not achieved by looking at the traceability matrix to ensure that every requirement is implemented. We need to go deeper that.

Tracking requirements and ensuring that they are comprehensively implemented belongs to quality assurance activities. In Chap. 6 we discussed the verification and validation of requirements. We also discussed the "how" of those two techniques. We included a bullet in Sect. 6.3.1, on Peer Reviews, which states "Ensure that the technical content is comprehensive." It is aimed at achieving this objective to ensure that the requirements are comprehensively implemented. Validation ensures that the implementation is indeed working and is defect free.

Verification and validation together ensure that the implementation of the requirements is comprehensive to the extent they are implemented in the software artifacts. The traceability matrix ensures that all requirements are implemented.

Quality assurance goes beyond the verification and validation techniques. Quality assurance includes the quality control activities (namely the verification and validation) as well as activities aimed at preventing defects and ensuring that the work is carried out comprehensively. These are discussed in Sect. 6.2, on "Quality Assurance in the context of Requirements Engineering and Management." Just to recap, the activities include defining and continuously improving an organizational process for carrying out various activities, staffing with the right resources well trained in their craft, diligent quality control, measurement and analysis of results, and project postmortem. The process includes procedures, standards, guidelines, checklists, formats and templates. All these play a vital role in ensuring that the requirements are comprehensively implemented throughout the software development life cycle. Quality control ensures that the implementation is right and is free from defects. Even the traceability matrix discussed in this chapter is also a format included in the organizational process assets from the formats and templates section.

Thus:

1. Tracing requirements using the traceability matrix ensures that all requirements are implemented.
2. The organizational process and diligent quality control ensures that all requirements are comprehensively implemented.

9.7 Requirements Reporting

It is one thing to carry out work conforming to a defined organizational process and it is yet another thing to ensure that work is carried out diligently, comprehensively and delivers results without defects. The difficulty increases proportionately with the distance you are separated from the level where work is carried out. It is more accentuated in large organizations. The people working on the job and the first level management would always be knowledgeable about the progress and quality of requirements implementation. It is the senior management and other stakeholders that would be in the dark unless we have a proper reporting system that keeps all stakeholders with the relevant information. Organizations normally use a weekly status reports in which the project management team prepares a comprehensive report on the project and communicates it to all stakeholders. This report would contain all aspects of the project under execution and provide the stakeholder with all that they need to know.

It is an exception rather than a rule to include the detailed progress achieved in the implementation of requirements in the weekly progress reports generated by the project management team. The weekly status reports contain the overall project progress, issues, items needing special attention from the senior management, change requests and so on but not the implementation of requirements. The change requests would be normally be reported as part of the weekly status report. The number of change requests received, rejected, accepted, completed, and under resolution would be normally reported. Sometimes, the effort spent on resolving the change requests and the corresponding impact on schedule would also find place in the weekly status report.

The argument for not making a special mention of the requirements implementation is that:

1. Their implementation is quantitatively monitored using the traceability matrix.
2. The qualitative aspect of implementation is monitored using the quality control activities.
3. The change request resolution is tracked using the change register and the weekly status report.
4. When you carefully consider, the progress of the project itself is the progress of implementation of requirements in the software artifacts (code as well as information artifacts) of the project. So, reporting project progress itself is progress of requirements implementation.

Therefore, it is felt that no separate reporting of requirements implementation is necessary in the project status report or monitoring activities.

9.8 Reconciliation of Requirements

It is necessary to finally reconcile the number of requirements originally accepted for implementation and the final number of requirements implemented at the completion of the project.

However, the first priority is to ensure that all customer requirements as amended by the accepted change requests are implemented comprehensively. This is accomplished by the acceptance testing. The main objective of acceptance testing is not to uncover lurking defects but to ensure that all the requirements are implemented in quantity and quality. In internal projects (that is the project deliverable is intended for use within the same organization), the acceptance testing is carried out by the end user department or the business analysts. In external projects (that is when the deliverable is intended to be used in another organization) the acceptance testing is carried out by the representatives of the customer. In product development projects (that is the deliverable is intended to be sold as a COTS product), the acceptance testing is carried out by a group of specially selected end users. This type of testing is known as "beta testing." They would not know the original set of requirements but they would test if the product meets their own requirements. By using a set of multiple testers, all the requirements would be covered by one tester or another.

Another activity that is carried out as part of reconciliation is a final inspection/ review of the traceability matrix. If any requirement is not traceable to its implementation, then, it has been missed. A corrective action would then be planned and implemented.

All in all, the reconciliation of requirements implementation is an important activity and is carried out in most organizations either explicitly or implicitly.

Chapter 10
Measurement and Metrics

10.1 Introduction

"What you can't measure, you can't understand, manage or improve" said a wise person. It explains succinctly the importance of measurement in the best possible manner. Measurement allows us to assess the performance in quantitative terms telling us where we stand comparatively. Without measurement, we can assess the performance only in subjective terms and it has been proven that subjective assessments can be erroneous. Measurement facilitates the following advantages:

1. It allows for an objective assessment of progress, and performance.
2. It allows us to compare our performance and benchmark it with other similar performances.
3. It allows us to set quantitative and fair targets to our resources and measure their performance against those quantitative targets.
4. It facilitates a *fair* system of rewards and punishments based on the quantitative data.
5. It facilitates computation of productivity and thus allows us an opportunity for productivity improvement and thereby, cost reduction.

All in all measurement is advantageous for the organization and it has come to be accepted by managements as essential for efficient management and improvement. Process standards like the ISO 9000 or the CMMI mandate measurement and analysis as well as using that quantitative data for organizational management.

10.2 Measurement and Metrics

In software development taxonomy, there is a plethora of terminology. For the same meaning, there are many terms floating around. No two standards agencies agree upon the same set of terms for a purpose. So, there is confusion among the

M. Chemuturi, *Requirements Engineering and Management for Software Development Projects*, DOI: 10.1007/978-1-4614-5377-2_10,
© Springer Science+Business Media New York 2013

practitioners with each claiming that their standard as the right one and all others are aberrations. Hence, if I do not define what I mean by the terms used in this book, it is possible that I may aggrieve/mislead some. Therefore, I would like to define the terms used in this book. Let us examine some popular definitions for popular terms used in the software development in the context of measurement and benchmarking.

Let us look at measurement first and then at metric.

Measurement

CMMI model document version 1.3 defines the term "measure" first. It defines measure as "a variable to which a value is assigned as a result of measurement". CMMI model document goes on to distinguish "measure" as "base measure" and as "derived measure." CMMI defines base measure as "A base measure is functionally independent of other measures." CMMI defines derived measure as "Measure that is defined as a function of two or more values of base measures." It defines "measurement" as "A set of operations to determine the value of a measure" and the measurement result as "a value determined by performing measurement."

From these definitions, we can, for our context, infer that:

1. Measurement is a process to determine the numerical value of some aspect of software development.
2. The result of measurement is a "measure" in numerical terms.
3. Base measure is the direct result of measurement.
4. We need to perform some mathematical transformations on one or more measures to obtain the "derived measure".

Metric—IEEE standard 610 Standard Glossary of Software Engineering Terminology defines metric as "A quantitative measure of the degree to which a system, component or process possesses a given attribute" and "quality metric" as "A quantitative measure of the degree to which an item possesses a given quality attribute" and "A function whose inputs are software data and whose output is a single numerical value that can be interpreted as the degree to which the software possesses a given quality attribute." CMMI model document version 1.3 used the term "metric" at a few places but left it undefined.

From the first IEEE definition of metric, we can see that it is very similar to the definition of base measure given by the CMMI model document. The second definition of IEEE for the "quality metric" is again, very similar to the CMMI definition of the "derived measure."

It would have been great if both (IEEE and CMMI) had agreed on a common set of terminology and definitions. It seems to me that the authors of CMMI wanted to move away from the oft used term "metric" and used the term "derived measure." Still, they used the term "metric" a few times in their model document version 1.3. That shows the popularity of the term "metric" in the software development fraternity.

For the purpose of this book, I am going to use the two terms, measurement, and metric.

Measurement is "A set of operations to determine the value of a measure" and measurement result is "a value determined by performing measurement"—the same definitions as of the CMMI.

A Metric is "A function whose inputs are software data and whose output is a single numerical value that can be interpreted as the degree to which the software possesses a given quality attribute." This is the IEEE definition. To clarify further, metric is a derived number from one or more measures and other metrics.

Now we have understood all the terms used in measurement n the software development industry.

Now let us look at benchmarking. IEEE standard 610 Standard Glossary of Software Engineering Terminology defines benchmark as "A standard against which measurements or comparisons can be made" and benchmarking as "A procedure, problem or test that can be used to compare systems of components to each other or to a standard."

Thus benchmark is an established quantitative value in the context of measurement and a standard in other contexts. Benchmarking is the process of establishing a benchmark. Unfortunately though, we do not have industry standard benchmarks in our software development industry. We have to establish our own benchmarks for our organization. Software industry follows the popular adage that "you are your worst enemy and competitor"!

Having put the terminology in its proper perspective, let us now look at the metrics that we can use in requirements engineering and management.

10.3 Metrics Relevant to Requirements Engineering and Management

Do we carry out measurement in software development? Yes, but perhaps not as diligently as in manufacturing or as one would wish. We measure the following attributes:

1. Effort spent by all resources using the organizational timesheets. Our measurement will be as good as our timesheet is. Effort is measured normally in person hours while in a few cases person days are also utilized. A good timesheet that can support effective measurement program would have:

   a. Employee id
   b. Cost center/Department id
   c. Project id
   d. Module id
   e. Software Development phase
   f. Software Development task
   g. Date
   h. Starting time
   i. Ending time

2. Size of artifacts using work register. The size would be different for information artifacts and code artifacts. Size of information artifacts is often measured in number of pages, number of requirements, number of classes and so on. Code artifacts are generally measured in LOC (Lines of Code), Function Points, Object Points or SSU (Software Size Units). Measuring software size forms part of the software estimation subject and is beyond the scope of this book. Interested readers may refer "Software Estimation: Best Practices, Tools and Techniques for Software Project Estimators" (2009) by Murali Chemuturi and published by J.Ross Publishing, Inc, USA for more information.
3. Defects uncovered in software artifacts, both the information artifacts and code artifacts. Defects are measured as integer numbers. They may also be classified as critical, major and minor or some such categorization. We often associate the origin of the defect with the measurement of defects. We obtain defect data from defect reports or defect resolution tools and the organizational defect definition which would define the class (critical, major or minor) and the origins.
4. We also make note of the dates on which activities are performed. We obtained the scheduled dates from the project schedule. We obtain the actual dates from the work register and the timesheets.
5. We also measure the change requests placed on the project to gauge their impact on the project and the stability requirements.

So, the primary measures are effort, size, defects, dates and changes. From these measures we derive a host of metrics for the project out of which a subset is relevant for the requirements engineering and management. In the following sections, we discuss those metrics that are relevant to requirements engineering and management.

We derive metrics in the following five classes:

1. Productivity Metrics
2. Change Request Metrics
3. Quality Metrics
4. Relative Effort Metrics
5. Schedule Metrics

Productivity metrics assist us in assessing the efficiency of handling the function.

Change request metrics allows us to assess the stability of requirements in the project.

Quality metrics assist us in understanding how well we performed the function and the level of quality in our deliverables.

Relative Effort metrics assist us in understanding the importance we are giving to the function in the overall scheme of software development.

Schedule metrics assist us in understanding how well we met the delivery commitments to our client.

Now let us look at each of them in a little greater detail.

10.3.1 Productivity Metrics

Productivity has multiple connotations. The one relevant here is that productivity is the rate of accomplishing a unit of work. It is the pace of working. It is the rate of delivering the assignment. While it is not pertinent to include more information on the concept of productivity in this book, I suggest that interested readers may refer to "Software Estimation: Best Practices, Tools and Techniques for Software Project Estimators" (by Murali Chemuturi and published by J.Ross Publishing, Inc, USA, 2009). The general formula for productivity is:

$$\textbf{Productivity} = \textbf{Outputs} / \textbf{Inputs}$$

Inputs are in person hours and outputs are in various units of work for software development work. It is expressed as so many person hours per unit size as in "10 person hours per function point". We can derive the productivity for the following activities of requirements engineering:

1. A gross productivity metric for all activities of requirements engineering put together. This would be useful for estimating the requirement of resources at the beginning of the project, especially for the senior management.
2. For each of the following activities:

    a. Elicitation and gathering
    b. Establishing the requirements

The formulas for each of these metrics are discussed below.

10.3.1.1 Gross Productivity Metric for Requirements Engineering

The formula is:

$$\textbf{GPM} = \textbf{E} \div \textbf{N}$$

Where:
GPM Gross Productivity Metric for Requirements Engineering
E Effort spent in person hours for all activities of requirements engineering including, elicitation, gathering, establishing, quality control, and change management
N Number of requirements as established in the requirements traceability matrix or URS

Data for this metric can be collated from:

1. Project timesheets in which the effort spent would be available.
2. Number of requirements can be collated from the traceability matrix or URS.

This metric can be derived only after the project is completed. It is easy to derive this metric because, we do not have to utilize special timesheets to extract effort data.

This metric is computed in most organizations and is utilized for estimating the resource requirements for the requirements engineering activity of the project. It is also utilized in cost estimation for the purposes of setting a budget for the project in the case of internal projects and for setting the price in the case of external projects to offer a quote against the RFP.

10.3.1.2 Productivity of Elicitation and Gathering

The formula for computing the productivity metric for elicitation and gathering is:

$$\textbf{PEG} = \textbf{EEG} \div \textbf{N}$$

Where:

PEG   Productivity for Elicitation and Gathering

EEG   Effort spent in person hours for all requirements engineering activities related to elicitation, and gathering

N   Number of requirements as established in the requirements traceability matrix or URS

Data for this metric can be collated from:

1. Project timesheets in which the effort spent would be available.
2. Number of requirements can be collated from the traceability matrix or URS.

This metric can be derived after the project requirements are established and are subjected to configuration management. We need not wait till the project is completed. It is not easy to derive this metric because, we do not get this data in the usual timesheets used in the industry. We have to utilize special timesheets to extract effort data.

Most organizations do not compute this metric because of the difficulty in obtaining the effort data. If computed, it gives better insight into how each of the major requirements engineering activities are consuming the total effort and thereby allows us an opportunity to effect focused improvement.

10.3.1.3 Productivity of Establishing the Requirements

This metric is computed to arrive at the productivity for establishing the requirements as URS and SRS. The activities, of documenting, verifying, validating, and implementing the feedback, are all included in the effort used for computing the metric.

The formula for this metric is:

$$\textbf{PER} = \textbf{EPR} \div \textbf{N}$$

Where:

PER Productivity for Establishing the Requirements

EPR Effort spent in person hours for all requirements engineering activities for establishing the requirements including, documenting, verifying, validating, implementing the feedback and approving the URS and SRS

N Number of requirements as established in the requirements traceability matrix or URS

Data for this metric can be collated from:

1. Project timesheets from which the effort spent would be available.
2. Number of requirements can be collated from the traceability matrix or URS.

This metric can be derived after the project requirements are established and are subjected to configuration management. We need not wait till the project is completed. It is not easy to derive this metric because, we do not get the effort data in the usual timesheets used in the industry. We have to utilize special timesheets to extract effort data required for computing this metric.

Most organizations do not compute this metric because of the difficulty in obtaining the effort data. If computed, it gives better insight into how each of the major requirements engineering activities are consuming the total effort and thereby allows us an opportunity to effect focused improvement.

Another aspect of this metric is that this activity is being downgraded from being "business analysis" to "technical writing" in many organizations by utilizing "technical writers." This is an emerging trend because this activity (documenting requirements) does not need fully qualified business analysts. It can be performed using the information collected during elicitation and gathering activities under the guidance of a business analyst. The advantage is reduction in the cost of requirement engineering as a whole. This practice is catching up in the industry. This metric helps us in determining the productivity of the technical writers and to set targets for them during project execution and better granularity during cost estimation.

10.3.2 Change Request Metrics

CRs are a reflection of the stability of the requirements. The argument is that if the requirements analysis is carried out diligently applying all necessary QA activities, CRs would not be needed. The CRR (Change Request Register) is the source of information for measuring the stability of requirements. These metrics are normally referred to as change request metrics or CR metrics.

10.3.2.1 Requirements Stability Metric

Requirements Stability Metric (RSM) metrics normally indicate the requirements stability. The following formula is used to compute requirements stability expressed as a percentage:

$$\textbf{RSM} = [(\textbf{Total no. of requirements} - \textbf{No. of CRs})/\textbf{Number of Requirements}] \times \textbf{100}$$

Another variant of this formula is:

$$\textbf{RSM} = (\textbf{No. of CRs}/\textbf{Number of Requirements}) \times \textbf{100}$$

where, RSM is the Requirements Stability Metric.

We derive this metric to gauge the stability of requirements. There is no industry benchmark for the requirements stability. We need to establish an organizational benchmark collating the data from past projects. We compare the metric for a just completed project with this benchmark to draw inferences and take corrective and preventive actions. We also use this metric for process improvement of requirements engineering.

The data for deriving this metric can be obtained from the project CRR and the traceability matrix or the URS. We can derive this metric only after the project is completed.

10.3.2.2 Relative Effort Spent on a Change Category

Another analysis that is carried out is the classification of changes into various categories so that the origin of changes can be determined and inferences drawn to see if any trend is emerging. Examples of scenarios are:

1. Suppose that the bulk of CRs are emanating from poor coding—then the organization will be alerted that additional training for coders is necessary.
2. Suppose that the bulk of CRs show that the understanding of customer requirements was not satisfactory, the organization will realize that Business Analysts need to be trained to be more effective in the process of requirements elicitation/gathering.
3. Suppose the bulk of CRs were due to defective design, then the organization would learn that software designers/architects should be improved.

In my opinion most categories of change requests could be reduced by one or more of the following suggestions:

1. Training to improve the skills of the personnel.
2. Better software development process.
3. Better tools and techniques.
4. Better standards and guidelines for coding, design, architecture and review.

5. Rigorous implementation of quality control.

Formula for computing the metric for a change category is:

RECC =   (effort spent on resolving CRS of a category/Total effort Spent on resolving all change requests) × 100

Where, RECC is the metric for Relative Effort spent on a specific Change Request Category.

The effort data for this metric can be collected from organizational timesheets and the CR data can be collated from the CRR. There is no industry benchmark for this metric. Therefore, we need to derive our own organizational benchmark. When the actual metric varies from the benchmark, we carry out an analysis to ascertain if the variance is due to random causes or assignable causes. If there are any assignable causes, we can draw inferences for improving the process to be implemented in future projects. We can derive this metric only after the project is completed.

10.3.3 Quality Metrics

Quality metrics assist us in finding the level of quality in our deliverables as well as to improve their quality level. Quality in a deliverable is the absence of the defects or conversely, presence of defects in the deliverables diminishes their quality. Therefore, we measure the level of quality using the defects discovered in our deliverables. We compute the below quality metrics.

10.3.3.1 Defect Injection Rate

When a person completes an artifact and hands it over for quality control activities, it is expected to be free from defects. But in practice, there would be some defects. These defects left inside the artifact by the author are referred to as "injected defects." This is a relative metric and we derive the number of defects against the number of requirements. We derive this metric from URS and the SRS. The formula is:

$$DIR = M \div N$$

Where:

DIR   Defect Injection Rate metric
M   Number of defects uncovered in quality control activities
N   Number of requirements as established in the requirements traceability matrix or the URS

We express this metric as "M" defects per "N" requirements. Let me illustrate this with an example.

Let us assume that there are 100 requirements in the traceability matrix and 5 defects were uncovered in all quality control activities that are attributable to requirements engineering. We express this as—**DIR is one defect for every 20 requirements**

Now what does it mean to us? No metric is meaningful without a comparable benchmark. While we do not have an industry benchmark for DIR, we have an ideal benchmark for delivered defects and that is 3 defects per one million opportunities. This is the ideal situation and we refer to the organization that achieved this level of quality as the "Six Sigma" level of quality organization. The level of quality is referred to as five sigma level if the delivered defects are 3 per one hundred thousand opportunities and four sigma if there are 3 delivered defects for ten thousand opportunities. Most professional organizations that have implemented a process to drive the organization would be between four sigma and five sigma levels at a minimum.

The philosophy of quality is to aim for zero-defects. We are now in the era of total quality management philosophy which states that we need to prevent error than to spend effort to uncover and fix it. Therefore, the DIR must be as close to the sigma level of the organization as possible. However, realizing that there would always be some defects left in the artifact by its author, we accept a variance of up to 20 % in the industry. That is if we are at four sigma level, then the DIR can be 3.6 defects per ten thousand opportunities or 36 defects per one hundred thousand opportunities or 360 defects per one million opportunities.

We can compute this metric after the project is completed as the defects may be uncovered during design, coding or testing in addition to requirements engineering that may be attributable to requirements stage.

10.3.3.2 Delivered Defect Density

We compute this metric for the overall project. We get this data only after the software product is put into production and is being used by the end users. The defect reports do come from the end users. Normally at this stage, it would be difficult to trace the origin of the defect. Another aspect is that unless all the accepted requirements of the end users are met, the product would not be accepted. Therefore, once the product is in production, we would not be getting defect reports whose origin lies in the requirements engineering stage of the software development. Therefore, I am not discussing this metric in this book. This metric is more relevant to software design and construction activities of the software development than to requirements engineering.

10.3.4 Relative Effort Metrics

Relative effort metrics help us in assessing the reasonableness of the importance given to activities. Absolute metrics do not tell us about the importance given to an activity. These metrics inform us of the relative importance being accorded to an activity in comparison with other activities. We compute the following metrics.

10.3.4.1 Importance to Requirements Engineering in the Project

This metric is computed using the formula:

$$\textbf{RRE} = (\textbf{ERE} \div \textbf{TE}) \times \textbf{100}$$

Where
RRE   Relative importance to Requirements Engineering in the project
ERE   Effort in person hours spent on all activities of Requirements Engineering
TE   Total effort in Person hours spent on all activities of the project

This metric is expressed as a percentage. We compare this percentage to the organizational benchmark and take appropriate corrective and preventive action as well as improve the process to control the variance in it. The data for this can be easily obtained from the organizational timesheets for the project.

While there is no established benchmark, 20 % is used in full life cycle (from requirements to acceptance testing and delivery) development projects to gauge the reasonableness of the amount of effort spent on requirements engineering. When a project's RRE exceeds the organizational benchmark, we carry out an analysis to ascertain what pitfalls caused this slide and draw lessons for future projects. If this metric falls below the organizational benchmark, we analyze the project to ascertain what best practices allowed the savings to improve our process.

But if we spent less time than the organizational benchmark, it could result in:

1. Higher DIR whose origin is in requirements engineering
2. Receiving more CRs

We can see a correlation between the effort spent on requirements engineering and the quality or stability of requirements. If the above two outcomes are absent, it establishes that we implemented some the best practices. So, this metric helps us in analyzing the requirements engineering activity and its impact on quality or CRs.

10.3.4.2 Quality Control of the URS and SRS

The main activities for ensuring that quality is in-built in the URS and SRS are verification consisting of peer review and managerial review and validation. The formula for computing this metric is::

$$RQC = (EQC \div ER) \times 100$$

Where:

RQC   Relative Effort metric for Quality Control of Requirements engineering activities

EQC   Effort spent in person hours for carrying out quality control activities namely, peer review, managerial review, validation, and implementing the feedback for both the URS and SRS

ER   Total effort spent in person hours on all requirements engineering activities

This metric is expressed as a percentage of the total effort on requirements engineering.

Data for this metric can be collated from project timesheets from which the effort spent would be available.

Quality control is an activity consuming resource but does not really adding any value to the product except uncovering the lurking defects. Therefore, the management is concerned if more than a reasonable amount of effort is spent on quality control activities.

We have to spend a minimum amount of effort on uncovering defects and ideally speaking, there should be only one, iteration for quality control activities. If there are more defects, it takes more iterations and consequently, consumes more effort for quality control. So, if the percentage is higher, it means that the quality of other requirements activities needs improvement. But how do we determine what is fair? One way is to establish the organizational benchmark for this metric and improve it over a period of time. While there is no approved industry benchmark for the metric, the industry uses 15 % as a heuristic as optimum percentage of time that can be spent on quality control of requirements engineering activities.

10.3.4.3 Relative Effort Spent on resolving CRs

Another metric normally derived is the amount of relative effort (expressed as a percentage) spent on resolving CRs using the below formula:

REC = (Total effort spent on resolving CRs / Total effort spent on the project) × 100

Where, REC is the Relative Effort spent on resolving CRs.

The data for computing this metric can be obtained from the organizational timesheets.

While there is no industry benchmark on how much percentage of time can be spent on resolving CRs, we can have a model metric for the organization. If a specific project overshoots this model metric, we may subject the CRs to deeper analysis to uncover the reason behind the high percentage of time spent on resolving the CRs. This metric helps us in understanding if the project effort is wasted on resolving preventable CRs.

10.3.5 Schedule Metrics

Effort metrics tell us the impact on the cost and thereby on the profit. While effort metrics are important for the organization, delivering the software, on schedule is much more important for the customer. Not achieving the benchmark effort metrics, would impact the cost and profit from the project but would not have strategic impact. Not achieving accepted schedules may even cause the project to be cancelled and thus has strategic impact on the organization. We compute the below schedule metrics.

10.3.5.1 Overall Schedule Metric

Here we compute the schedule metric for the schedule of the completion of requirements stage in the software development signified by the establishment of URS and the SRS and subjecting them to the rigor of configuration management. This metric is expressed as a percentage. The formula for deriving this metric is:

SMRE = [(**Number of calendar days actually taken − Number of calendar days scheduled**) ÷ **Number of calendar days scheduled**] × **100**

where SMRE is the Schedule Metric for Requirements Engineering

This metric is positive when we overshoot the schedule, that is, the delivery is delayed and negative when we beat the schedule and delivered before the scheduled delivery date.

While there is no industry wide benchmark for this metric, any amount of delay is frowned upon.

Another important question is which scheduled date is to be considered—is it the original schedule or the amended schedule?

From the standpoint of organizational efficiency, we need to take the original schedule into consideration. It may not always be to our liking. We may use the amended schedule if we wish to present an attractive set of metrics.

The data for deriving this metric can be obtained from the project schedule and the work register. This metric can be derived as soon as the URS and SRS are subjected to the rigor of the configuration management.

This metric assists us in learning about our capability to meet the schedules and improve them if necessary.

10.3.5.2 Individual Schedule Metrics

These metrics are just the same as the overall schedule metric except that these are computed for each of the persons involved in the requirements engineering activity. Why do we need these? By deriving this metric for each of the staff, we can ascertain who is delivering the best performance and who the bottleneck in the delivery chain is.

10.4 Summary of Metrics

The Table 10.1 summarizes the metrics as relevant to requirements engineering activity.

Table 10.1 Metrics relevant to requirements engineering

Metric	Purpose
Productivity metrics	To ascertain the efficiency of our performance
Gross productivity metric	1. Assists in cost estimation
	2. Assists in estimating the resource requirements
Productivity for elicitation and gathering	To assess the efficiency of elicitation and gathering activity
Productivity for establishment of requirements	To assess the efficiency of the establishment activity of requirements
Resolving change request Metrics	To assess the impact of changes on the project
Requirements stability metric	To assess the volatility of requirements
Relative effort spent on a change request category	To identify the change category which is causing more change requests and thereby identify weak areas of requirements engineering activities
Quality metrics	To assess the levels of quality of the deliverables of the requirements engineering
Defect injection rate	To assess how well the requirements engineering is being carried out in the project in the first iteration
Relative effort metrics	To assess the importance accorded to requirements activity in the overall scheme of the project
Relative importance to requirements engineering in the project	To assess the importance received by the requirements engineering activity in the overall project
Quality control of URS and SRS	To assess if we spent reasonable amount of time in quality control
Relative effort spent on resolving CRs	To assess if the time spent on resolving CRs is reasonable
Schedule metrics	To assess how well we are meeting the accepted schedules
Overall schedule metric	To assess if we have met the schedule of the final delivery of requirements engineering activity
Individual schedule metrics	To assess how well individual resources are meeting their accepted schedules

Chapter 11
Roles and Responsibilities in REM

11.1 Introduction

To achieve the objective of handling the activity of requirements engineering and management efficiently and effectively, we need to define the roles and responsibilities of all the agencies involved in the activity. Once defined, the concerned individuals ought to be trained to perform their roles in the letter and the spirit of the definition. We also should have checks and balances built into the system to ensure that the defined process is implemented effectively in the organization as well as ways to trigger improvements thereof.

There are two important agencies that should collaborate and function effectively for any activity including requirements engineering and management, to be successful. One is the organization itself and the other is the set of individuals involved with the function. We will be discussing the roles of these two agencies in greater detail in the following sections.

11.2 Role of the Organization

The main role of the organization is to provide an environment conducive to producing excellent results in any endeavor by the human resources. Organizations facilitate and the individuals perform. What are the elements of an organizational environment that have to be designed so as to facilitating excellent performance? Here are the elements:

1. Organization
2. Staff
3. Process
4. Quality Assurance

M. Chemuturi, *Requirements Engineering and Management for Software Development Projects*, DOI: 10.1007/978-1-4614-5377-2_11,
© Springer Science+Business Media New York 2013

5. Training
6. Recognition and rewards

    Let us discuss each of them in detail.

11.2.1 Organization

By organization, I mean the arrangement of various departments in the organization. An organization is arranged into various departments each having its own set of responsibilities and a concomitant authority. When the departments are properly organized, the departments work in a close-knit manner putting shoulder to shoulder and support each other to produce synergetic results that are better than the sum of individual efforts. When the departments are poorly organized, departments do not support each other; do not communicate with each other; they necessitate coordination; there will be too many meetings to resolve issues; escalation becomes too frequent; the results would be less than the sum of individual efforts; and the quality of the deliverables would be impacted. Therefore, the organization ought to be diligent in organizing the departments. Each department must have a clear role and a set of deliverables with concomitant authority to accomplish results.

    While the subject of organizing in general is not in the scope of this book, we need to recognize that it plays a vital role in the efficient functioning of the organization. Another important aspect is that the organization must recognize the importance of the activity of requirements engineering and management and provide a place for it in the organization. Most organizations do not have a specialist department for requirements engineering. Business analysts or systems analysts are normally part of the project team that delivers the software and they report to the project manager. If the organization has multiple projects in execution, there will be business analysts spread across many projects. But there is no central department that owns these analysts to focus on their development. Business analysts or systems analysts are usually part of the software development pool in software development organizations and are part of the IS department in other organizations.

    Slowly but surely, the field of requirements engineering and management is severing itself from programmers. A few years ago, senior programmers were carrying out this activity. But now, it is no more so. It is evolving itself into a separate specialty. While I do not advocate a separate department for the activity of requirements engineering and management, I do advocate a support group for this activity separate from the project. When the business analyst needs a consultation, there ought to be a place from where to obtain it from within the organization. Organizations ought to give consideration to this specialty and provide a support group for requirements engineering and management. The organizations specializing in software development as their main source of revenue ought to dedicate a department to support this activity.

Another important support to be provided by the organization is the knowledge repository. Each organization would usually have a knowledge repository, it is doubtful if it contains material concerning requirements engineering and management. The knowledge repository ought to have a section on requirements engineering and management which must be updated with the latest developments in the field.

11.2.2 Staff

For any department in the organization to deliver effective results, it needs competent staff and the activity of requirements engineering and management is no exception. It needs qualified staff, trained in the field of requirements engineering and management and mentored on the job to be able to deliver effective results.

Deciding on the qualifications needed for carrying out requirements engineering work, it must be noted that it is not the same for every type of project. Different types of projects need differently qualified persons. Let us look at project types and the needed qualifications one by one.

11.2.2.1 Projects Implementing a COTS Product

There are a variety of excellent COTS products with built-in industry best practices, available at a cost that is lower than the cost of custom-software development for comparable functionality. These are available in the fields of ERP (Enterprise Resources Planning), SCM (Supply Chain Management), CRM (Customer Relationship Management), EDI (Electronic Data Interchange). EAI (Enterprise Applications Integration), Financials, Banking, Manufacturing, credit card processing, telecom billing and so on. Most of them can be implemented as they are or with some extension or customization. In these types of projects, the main activity of requirements engineering would be mapping the functionality available in the COTS product with the requirements of the organization; bringing out a gaps document; carrying out the acceptance testing; and handhold the customer during commissioning and rolling out of the product. This activity needs functional specialists trained on the COTS product under implementation. To implement a financials COTS product, we need persons qualified in accounting with experience in handling accounting function. To implement a HR COTS product, we need individuals qualified in the HR subject with some experience in a HR department. To implement, SCM COTS product, we need individuals qualified in the material management/supply chain management subject with experience in a materials management department. Usually, it is common to take individuals with MBA

(Master of Business Administration) or an equivalent qualification with some experience in their respective functions and then train them on the COTS product before putting them on the job. Can a newcomer be used in these positions? Strictly speaking,—no. But, the individual can be taken a s a trainee and mentored on the job by senior staff for some time before putting them on the job independently.

So in this type of project, it is better to use persons with professional qualifications in the respective field with some field experience and training on the respective COTS product.

11.2.2.2 Full Life Cycle Software Development Projects

Full life cycle software development projects begin with requirements analysis and end with acceptance testing at a minimum. Project acquisition and implementation are also included in internal projects. This project type is perhaps the oldest in all software development projects. The practice from the beginning has been to utilize senior programmers for requirements engineering work which was referred to as "systems analysis" in those days and until recently.

After the advent of ERP and its implementation projects, functional specialists made an entry into software development projects. Managements have been quick to grasp the merits of using functional specialists in requirements engineering and started using them in full life cycle development projects with much better results than when senior programmers were used. So, now the trend is to use functional specialists for requirements engineering in these full life cycle projects even though the practice of systems analysis is not yet extinct.

So, just as in COTS product implementation projects, business analysts are being utilized more and more frequently for requirements engineering activity. Presently, MBAs are being recruited with some experience in functional domains and are trained in the requirements engineering specialty for use as business analysts in full life cycle software development projects. The practice of recruiting fresh graduates from business schools is also picking up. However, these individuals need to be mentored on the job before giving them independent charge for a new project.

11.2.2.3 Testing Projects

Just a few years ago, testing was just an appendage of software development. Software developers used to self-test and certify the product for use. Independent testers were unheard of! Gone are those days and IV and V (Independent Verification and Validation) has become an essential part of software development. With the advent of web based projects the need for testing the software products increased exponentially. Additionally, the need for certification of COTS products for use is becoming rather mandatory to prove that they are functionally adequate and are safe for use. Because of the complexity involved in testing the software for use on the Internet, most organizations are outsourcing the testing activity, so much so, now there are many organizations specializing only in testing.

Requirements engineering in these projects is entirely different from other projects. In other projects, the focus is on getting the end users to define their needs comprehensively and document them in such way that software engineers can develop the desired product. In testing projects, we need to ensure that the product works as specified. The specifications are already available in URS, SRS and possibly the design document. The requirements besides what is contained in the URS and the SRS are the timelines that need to be adhered to for completing the testing.

Then comes the designing of the test strategy, and developing the test cases. Testers would carry out the testing using the test cases.

The need is to find out how the end users would use the product and simulate that usage during testing to ensure that it works and works without defects. The test case designer, if not the tester, ought to know the difference between the right result and the wrong result. Therefore, the test case design is normally carried out by functional specialists because they are best positioned to know if the product covers the functionality comprehensively.

In some testing projects, there may not be any documentation in which case, there is a need to elicit and gather testing requirements from the end users. If it becomes necessary, then the requirements engineering is similar to that of full life cycle software development projects.

Most organizations use functional specialists for requirements engineering and test case design in testing projects. Fresh business graduates are also being used with training on software testing. I think that using functional specialists is the right approach.

11.2.2.4 Conversion/Porting/Migration Projects

In these projects, there already exists a working software product which needs to either be converted or ported to another platform or migrated to a newer version of the existing platform.

Conversion projects are like Y2 K or Euro conversion projects. The software product is already working but needs to be converted to accommodate the year 2000 or Euro functionality.

When an application is moved from one hardware platform to another but using the same programming language, we refer to those projects as porting projects. The need is to adjust the code of the product to handle the specific differences between both the platforms.

Migration projects involve shifting the application on the same hardware platform but on to its next version of the software platform. The newer version may have a few differences of syntax and may have some extra facilities too. So the application needs adjustment to be able to handle the differences and make use of the additional facilities provided by the newer version.

In these projects, the product already exists and there is no need to go to the end users again. Perhaps, the documentation and test plans also may exist in some cases. The important activity is to locate the adjustment needed and effect it. Everything is technical in nature as code verification is needed to locate the differences.

Requirements engineering in these projects involves locating the adjustments necessary to the existing code. It is best achieved by senior programmers. Functional specialists are poorly equipped to come out with the necessary code adjustments. Therefore, it is best to use senior programmers for carrying out requirements engineering in these projects.

11.2.2.5 Partial Life Cycle Projects

A partial life cycle project can be any combination of the phases of the software development life cycle. Normally the product implementation, requirements engineering and the construction phases are outsourced. Software design is rarely outsourced. Product implementation is one of the most outsourced projects. We had examined the scenario of product implementation in one of the above sections. That leaves us requirements engineering and construction.

For the projects involving requirements engineering, the only requirements to be gathered are the timelines and the cost during project acquisition. These can be handled by the marketing personnel. The project would be executed by business analysts. Needless to say perhaps, these need to be functional specialists.

For construction projects in which the project involves developing the code and conducting quality assurance activities conforming to a design document which is usually supplied by the outsourcer. Here the need to understand the technical aspects is more important than understanding the business functionality. Therefore, senior programmers or the systems analysts are best positioned to carry out the requirements engineering activity, which involves understanding the technical aspects. In fact, there is no requirements engineering activity except to understand the design document and deliver code adhering to that document.

Sometimes, only the requirements documents may be supplied and software design may also be included in the construction projects. In those cases too, the requirements are already established. So, there is no need for functional specialists. The technical persons can handle the remaining little portion of the requirements engineering activity.

11.2.2.6 Projects Developing Real Time Software

Real time software is used to control machines including cars, aero planes, CNC (Computer Numerically Controlled) machines, rockets and many others. In fact, there is hardly any machine in the present day that is not software controlled. In this software, there are no business functions. There are only technical functions.

Table 11.1 Persons appropriate for requirements engineering activity project-wise

Project type	Persons appropriate for requirements engineering activity
Implementation of COTS project	Functional specialists with training on the respective COTS product
Full life cycle software development projects	Functional specialists
Testing projects	Functional specialists
Conversion/porting/migration projects	Senior programmers
Partial life cycle projects—requirements engineering	Functional specialists
Partial life cycle projects—software design, construction and testing	Senior programmers
Partial life cycle projects—construction and testing	Senior programmers
Real time software projects	Senior programmers

Normally, the real time software is developed by technical people with engineering qualification especially from electronics engineering or similar background. The requirements are all technical in nature.

The requirements engineering activity is therefore handled by technical persons especially by senior programmers who handled similar projects earlier.

Table 11.1 enumerates the appropriate individuals to handle requirements engineering activities.

11.2.3 Process

Another important aspect for which the organization owns the responsibility is definition and continuous improvement of an appropriate process for the requirements engineering activity. While quality control activities uncover defects and facilitate their correction, a well-defined process would ensure that quality is built into the deliverables right during the engineering stage itself, and reduces the effort spent on quality control besides facilitating improvement in the efficiency all around.

An organizational process consists of a network of procedures, standards, guidelines, formats, templates and checklists.

A procedure consists of:

1. Step by step instructions on how to accomplish a specific task
2. Instructions to ensure that quality is built into the deliverable
3. Suggestions to prevent defects in the deliverable
4. Suggestions to ensure efficiency in the utilization of resources
5. The list of suggested quality control activities to ensure that the deliverable is defect-free
6. References to associated standards, guidelines, formats, templates and checklists that can be used in performing the task

Procedures help a new entrant to perform on par with an experienced resource and an experienced resource to perform at the peak efficiency.

A standard is something that is established by authority for use in the performance of an activity. In an organizations, standard is a document that consists of the selected alternative for use by the organizational resources from among various available alternatives. A standard is a restriction because it restricts the freedom of people in selecting the alternative of their choice. But by spending effort and resources to evaluate all the available alternatives beforehand to select the optimal alternative for use within the organization, it helps resources by reducing their effort to make a wise decision in every project. It also ensures a uniform level of quality in the deliverables of all projects. It is because of standardization the cost of various products including cars have come to affordable levels.

A guideline is akin to a standard except that the standard is prescriptive while a guideline is suggestive. A guideline guides a person in selecting a suitable alternative and various other aspects associated with it but stops short of prescribing a specific alternative.

A format is a document with a prescribed organization of information that aids in capturing information comprehensively in the first place and to aid in understanding the information contained in the document by the reader. It delineates the documents into sections, tables and frames to capture and present information efficiently.

A template document is similar to a format document except that it contains explanation in each cell about what information to be placed in the cell. It would also contain selection lists so that instead of entering information some boxes can be checked.

A checklist is a list of items usually neglected in a deliverable document. A checklist contains a number of items against which, we need to enter either "yes" or "no". Usually, all items must have a "yes" marked against it. All items that are marked "no" need to be revisited again. A checklist is used by the authors to ensure that the deliverable is comprehensive and by the quality control persons to ensure that nothing is left out of the deliverable.

It is the responsibility of the organization to define a process that is appropriate for the organization and implement it. As change is constant in modern organizations, the organization also needs to institute a mechanism to review the process periodically and improve it to ensure that the process is relevant even in the changed conditions. Normally each software development organization would have a Software Process Group which would initially define the process, then implement it and capture suggestions for improvement. The improvement suggestions are evaluated and appropriate ones are picked up and dovetailed into the process periodically.

A well-defined process for requirements engineering is a prime requisite for ensuring that it is carried out efficiently in the organization and it is the responsibility of the organization performed through the organizational process group.

11.2.4 Quality Assurance

Quality assurance includes both defect prevention and defect detection. Defect prevention is ensured using the defined organizational process. Quality control needs to be performed at the project level to uncover all lurking defects so that they can be fixed.

The quality control in an organization is like police in a town. Existence of police cannot prevent a determined criminal but would deter any criminal with lesser determination from committing a crime. Besides, the police would catch the criminal, well almost all criminals. So is quality control; it cannot prevent all the defects from being injected but it would prevent most and trap most of the remaining ones so that we deliver a near defect-free deliverable.

Presently, many software development organizations do not have a robust quality control department. In organizations that develop software for in-house (within the organization) use, independent quality control may itself be totally absent. This is not conducive to delivering good quality outputs.

It is the organization's responsibility to establish a robust quality assurance department and ensure that rigorous quality control is carried out on all the deliverables including the requirements engineering deliverables. Quality control activities may not prevent injection of defects but uncovers them and ensures that the final deliverable is as defect free as humanly possible.

11.2.5 Training

Having qualified staff is essential to perform requirements engineering activities in the organization efficiently. But periodic training to update and hone their skills is essential if we need to continue performing at the highest level. These days, a significant amount of research is being carried out in every field of human endeavor and so is the case with the field of requirements engineering. The way requirements engineering activities were performed in the 1970s is vastly different from the way they are now performed. If we do not train and keep our staff at the cutting edge of developments, they will soon become obsolete very quickly, especially in these days of fast obsolescence.

A training department in the organization would go a long way in ensuring that all the staff is adequately trained. The training department ought to perform the following activities:

1. Maintain the skill database of all employees of the organization
2. Organize a knowledge repository to facilitate self-study to update knowledge
3. Assess the training needs of the staff at regular intervals and analyze the gaps in the skills available and the skills needed
4. Draw up training plans to bridge the skill gaps uncovered in the above analysis

5. Maintain a databank of various public training programs available in the vicinity and the external faculty that can be called upon in times of need to be able to conduct various training programs
6. Organize suitable training programs to update the skills and knowledge of the resources
7. Evaluate the efficacy of the training programs conducted within and without the organization
8. Perform all other functions necessary to keep the employees on the edge of technology of their respective specializations

Therefore, a training department would go a long way in ensuring that all employees possess the knowledge necessary to perform their activities at the peak efficiency.

The training department not only organizes training programs but also deputes staff to attend various seminars to learn about the latest research and developments taking place in the field.

It is the responsibility of the organization through its training department to provide necessary training to its staff and maintain the skills of its staff up to date.

11.2.6 Recognition and Rewards

People are typical. When they are out of job, they promise to work to their full capacity but once they get one, they, or rather most of them, resort to penalty-avoidance level of working. This is well recognized in the industry and research has been carried out about how to motivate employees to give their best to the organization. It is now widely recognized that recognition and rewards are the best way to motivate employees toward attaining and contributing their full potential to the organization.

We need to institute a system of recognition and rewards to the individuals performing the requirements engineering activity. These must be fair and the method of selecting the individuals for rewards must be transparent and based on quantitative data. In some cases a certain individual always performs best and comes first. Organizations commit the mistake of giving the reward to the same person consecutively. This makes all others feel completely demotivated. If a person is consistently performing at the top of the pack, it is better to promote him/her to the next level. It is better to give the rewards in such a manner that everyone should feel that he/she has a fair chance of getting it. That way, the individual would be motivated to try for the reward.

Should the reward always be financial in nature? Not necessarily. It can be just a mention in a large gathering, or a certificate or some such other thing. In some organizations, they simply give a star to be pinned on their lapel. The more stars one has pinned on the shirt, the more recognized that person would be in organizational gatherings.

Whatever method the organization chooses, for recognizing and rewarding, it is necessary to have a system of recognition and rewards to motivate in the organization to motivate the staff toward higher level of performance.

All of those above activities are the responsibility of the organization. All these activities create an environment that is conducive to producing great results and delivering quality outputs. Once the environment is in place, the individuals can excel in that environment. One cannot excel in a poor environment.

11.3   Role of the Individuals

An organization has the onus of creating the right environment in which individuals can excel. But ultimately it is the individuals that perform the function and deliver the results. We have the following individual roles in the context of requirements engineering and management in software development projects:

1. Business/System Analysts
2. Quality Control
3. Project Manager
4. Process definition and improvement group
5. Senior management

Let us discuss the role of each of these individuals.

11.3.1   Business/System Analysts

These are the people who bear the brunt of carrying out the requirements engineering work. They bear the primary responsibility for the deliverables within accepted schedule and at the best possible quality. They carry out the requirements elicitation, gathering, analysis, establishment of requirements preparing the traceability matrix initially and the acceptance testing finally to ensure that the final product meets all the requirements effectively. The rest of the individuals either perform quality control on their deliverables or supervise (project manager/ leader) them or work under (technical writers) them. What are the responsibilities of this set of people? Here they are:

1. Take ownership of the requirements engineering activity in a project or as directed by the organizational management
2. Carry out all requirements activities diligently, accurately and efficiently and on time
3. Establish project requirements on time and as defect-free as possible
4. Keep their knowledge up to date through organization sponsored programs and self-study

5. Assist the organization in the definition of the organizational process for requirements engineering as well as continuously improve it to keep it up to date plowing back the experience gathered while executing the projects
6. Assist the organization in setting up a knowledge repository for the subject of requirements engineering
7. Assist the organization in the recruitment of personnel to fill the positions of business analysis through recruitment and selection
8. Assist the organization in training the new recruits for induction into projects to carry out the work of requirements engineering
9. Evaluate tools and techniques for carrying out requirements engineering and to recommend appropriate ones for acquisition to improve the efficiency of the activity
10. Contribute to the profession in general to extend the frontiers of knowledge of requirements engineering
11. Any other organization specific activity.

One aspect worth noting here is about the career path for business analysts. Earlier, the systems analysts were progressing towards leading a project and then on to project management. Because the systems analysts grew from programming work, leading and managing projects was easy and presumed to be natural. But when we recruit functional specialists and assign them to carry out requirements engineering activities, promoting them to project management seems risky. But organizations have been promoting business analysts to the role of project manager and they seem to be performing ably. I would suggest that re-orientation training be imparted before promoting a business analyst to a project manager. This would help the business analyst to appreciate the technical issues with better clarity.

11.3.2 Quality Control

We have explained in detail the quality control activities as applicable to requirements engineering. The persons entrusted with the quality control activities ought to perform their activities diligently so that all and any lurking defects are uncovered and passed on for fixing. They ought to err on the positive side, that is, finding more errors even if they are perceived to be frivolous rather than allowing a defective deliverable to slip through. Their responsibilities are:

1. Take ownership of the quality of the deliverable in a software project
2. Carry out the quality control activities of verification and validation diligently and uncover all defects lurking inside the deliverable
3. Raise a defect report for all quality control activities carried out and follow through until all defects are satisfactorily resolved
4. Assist the project management in the project postmortem and discuss all defects so that they can be prevented in the future projects

5. Assist the organization in training new recruits in quality control so that they can be inducted into the project quality assurance quickly
6. Contribute to the organizational initiatives in process definition and improvement for quality control activities

Quality control is a thankless job and is perceived to be negative by other individuals of the organization as they have the onerous task of pointing fingers at other's errors. Therefore, management ought to support quality control personnel to ensure the quality in the deliverables by finding and removing defects from them.

Many organizations have the quality control activities performed by peers, that is, by other business analysts/systems analysts working on projects. The management ought to ensure that uncovering defects should not be hampered by "you scratch my back and I will scratch yours" syndrome. The organization should inculcate a culture in which quality control is perceived as an exercise in removing defects rather than an exercise in pointing fingers.

11.3.3 Project Manager

A project Manager (PM) is like the thread in a garland of flowers. A PM integrates the efforts of all individuals into a finished product that is as defect-free as humanly possible; meets the requirements of the customer and is delivered on time, within the sanctioned budget or cost agreed upon.

A PM is the immediate layer over the business analysts/systems analysts and has the potential to impact the requirements engineering activity to a great extent both positively and negatively. A PM provides facilitation to the business analysts/systems analysts so that they can perform and excel. A PM is the one to ensure implementation of all the organizational processes, standards, guidelines, formats, templates and checklists. If the PM is diligent, the organizational processes would be implemented effectively.

The responsibilities of a PM in the context of requirements engineering are:

1. Plan the activities of requirements engineering along with all other project execution activities
2. Schedule the activities pertaining to requirements engineering giving them due consideration and allocating it all the necessary resources of people, time, and money
3. Request and obtain required number of qualified and trained resources to carry out the activities effectively and efficiently
4. Allocate the work of all requirements engineering activities to appropriate resources on time, set fair targets and de-allocate appropriately
5. Appraise the performance of the resources fairly
6. Assist the organization in the recognition of performance and giving rewards to deserving resources without any prejudice or bias
7. Take all actions necessary to motivate the resources and keep the team morale at the highest possible level

8. Spare the time of the resources to assist the organization in the recruitment, training and process definition/improvement activities as required by the organization
9. Administer disciplinary actions as necessitated fairly
10. Enforce schedules so that all deliveries are effected on time

A project managed well by a qualified and experienced PM would deliver better quality deliverables than a project managed by poorly qualified/experienced project manager. When the PM is a bit less than desirable in terms of experience/qualifications, it would be in the organization's best interest that a senior manager closely oversees the project execution, while mentoring the PM toward better performance.

11.3.4 Process Definition and Improvement Group

Professional organizations do normally have a process for carrying every activity in the organization. They do normally earmark a set of people to champion definition and improvement of the process. This core group champions and facilitates the process definition and improvement. The actual definition is carried out by experts drawn from the functional groups within the organization or sometimes when the right expertise is not available, experts may be drawn from outside the organization. The responsibilities of this core group are:

1. Champion the definition and improvement of process in the organization
2. Identify the right experts from within or without the organization and provide them facilities for the process definition, review the feedback/change requests placed for process improvement, and dovetail the feedback/change requests into the process
3. Coordinate the quality control of the process artifacts and obtain approvals from concerned authorities for implementing the process in the organization
4. Pilot and implement the process in the organization
5. Take ownership of all process assets in the organization and safeguard them against unauthorized changes
6. Take all actions necessary to internalize the process in the organization
7. Provide handholding assistance to organizational resources in the right implementation of the defined process
8. Institute appropriate mechanisms to receive, analyze and implement feedback/change requests received from various agencies regarding the process
9. Periodically review all approved feedback and arrange to implement the feedback, pilot it and roll out the revised process in the organization
10. Coordinate with certification agencies and obtain desired certifications for the organization
11. Coordinate any other activity relevant to process definition and improvement in the organization.

11.3.5 Senior Management

It is impossible to do anything in an organization without the support of its senior management. Requirements engineering and management is no exception. The role of the senior management in ensuring that the requirements engineering activity and management is carried out efficiently and effectively is:

1. To provide adequate resources to carry out requirements engineering activity in the organization
2. Provide facilities and funds to impart the required training to keep the business analysts/systems analysts at the leading edge of technology and practice
3. To facilitate definition and improvement of organizational processes
4. Accord approval to all process artifacts for implementation in the organization after due managerial review
5. Periodically review the performance of the practice of requirements engineering and management within the organization and take necessary corrective and preventive actions to keep the practice at its best
6. Appraise the persons involved in the requirements engineering and management in the organization and accord recognition and rewards periodically
7. Enforce discipline on the resources whenever it becomes necessary to eliminate deadwood and keep the team motivated
8. Take all necessary actions to motivate the resources and keep the morale very high in the organization in general and in the resources involved with the requirements engineering activity in particular
9. Any other activities necessary to ensure that the practice is carried out efficiently and effectively in the organization
10. In fact, all the responsibilities specified earlier in the section on the role of the organization are the responsibility of the senior management.

11.4 Final Words

If any organizational endeavor has to succeed, the collaboration between the organization and the concerned individuals is essential. The Organization is represented by the senior management. The Organization creates an environment conducive to pursue excellence and produce desired results. Individuals utilize that environment and produce desired results and move the organization towards achieving excellence.

Chapter 12
Requirements Management Through SDLC

12.1 Introduction

While establishment of requirements is among the first activities in the software development life cycle, requirements need to be managed through the SDLC (Software Development Life Cycle) to ensure that all the customer requirements are included in the final deliverable. To do so, we need to include the additional requirements received through the change requests, delete the requirements that are eliminated through change requests and modify the requirements modified through the change requests.

But, the industry does not have a single standardized SDLC and in fact, there are multiple SDLCs. Here are some of the popular SDLCs:

1. Water fall model
2. Spiral model
3. RAD and JAD
4. Iterative model
5. Agile methods

    a. XP
    b. Scrum
    c. RUP agile version
    d. Many more

There are many SDLCs out there in the software development industry. While a detailed discussion of all the life cycles out there is out of scope for this book, the following phases are commonly part of any SDLC:

1. Requirements
2. Software Design
3. Construction
4. Testing

M. Chemuturi, *Requirements Engineering and Management for Software Development Projects*, DOI: 10.1007/978-1-4614-5377-2_12, © Springer Science+Business Media New York 2013

In addition, we need to consider the pre-project phase as well as the post-delivery installation and commissioning phase as both have an impact on the requirements management of the project. However, each of the SDLCs implement these phases in their own way and the artifacts recommended also vary significantly. Still all four phases are used albeit with differences in all SDLCs. Let us now discuss how the requirements are managed in each of these phases.

12.2 Pre-Project Phase

While the pre-project phase is not part of SDLC, it is relevant to requirements management. The project is a sequel to fulfill an existing requirement. In this phase, the requirement for software is recognized and established to the extent that the investment to fulfill that requirement is justified and approved. In this phase, the following activities are carried out:

1. Recognition of the need—the need for computerized information processing or the need to upgrade the existing system is recognized by a concerned executive in the organization. Then the need is articulated to the management so that attention is focused on that need and investigation is initiated to ascertain the efficacy of the need. In some organizations, including software development organizations, BPO (Business Process Organizations), and others, the IT infrastructure, including application software is part of the original investment itself. But upgrading the IT infrastructure in those organizations again has to be recognized and perhaps the marketing department or the strategic planning department or the senior management itself recognizes the need. This activity recognizes the presence of a need for IT infrastructure, including application software, and approves further investigation to establish or reject the need.
2. Investigation of the need to approve feasibility study—senior management would consider the need recognized by the concerned executive and approve expenditure toward further investigation of the need or reject the proposal. Once approved, a feasibility study is commissioned to establish the requirement or reject it.
3. A feasibility study is used to establish the requirement and draw up the project specifications—it is conducted to justify or reject investment in the proposed computerized information processing need. A feasibility study investigates the need, captures, the volumes, resources (people, equipment, money, and duration) needed to fulfill the need, carry out cost-benefit analysis and draw up project specifications. The deliverable of the feasibility study would be a feasibility report.
4. Approval of investment—The feasibility report would be considered by senior management and depending on the availability of funds, approves the project. This will spur the project into action.

5. Project initiation—Once the funds are approved and a budget is sanctioned, in principle, the project is initiated. The project initiation activities include

   a. Identification and allocation of the project manager.
   b. Project planning.
   c. Setting up the development environment including allocation of workstations, servers, connectivity and seating facility.
   d. Allocation of team members.
   e. Project kickoff meeting and handover the project to the project manager.

In the context of requirements management, the pre-project phase concludes with the project initiation. Establishment of the need for the application software is accomplished in the pre-project phase.

12.3 RM in Requirements Phase of SDLC

In the requirements phase, primarily, we establish the requirements for the project. Chapters 2–5 of this book, deal with this topic in detail. Chapter 6 deals with ensuring quality in the established requirements. In view of the topic being dealt with, in detail, in the preceding chapters, there is no point in briefly repeating it here.

A requirements traceability matrix is crated and initiated in this phase apart from establishing the project requirements in detail. This matrix would be updated in all the subsequent phases.

12.4 Software Design

The established requirements are the input for the software design phase. But carrying out design could discover opportunities to provide extra functionality without too much additional cost or sometimes without any additional financial burden. Sometimes, we may modify some of the requirements to suit the available technology. For example, a printed hardcopy report may be replaced by a screen based enquiry. Another opportunity may be to send an email in token of receipt instead of a printed and signed receipt. We need to remember that end users are not exposed to all the facilities available in the computers or the capabilities of the modern computers. They would not be able to specify in such a way that the capabilities of the computers and software development platforms are exploited to their fullest potential. Therefore, during the design phase, we ought to look for opportunities for improvement in the functionality which is beneficial to the end users or the organizational management. Another aspect of requirements management we need to handle in the design phase is the change requests that are placed by end users or any other stakeholder. We need to implement all the change requests received thus far, in design phase itself so that the software construction team would not be hindered by design changes during construction of the software.

The Requirements traceability matrix would be updated including the references from the design document against each of the requirements. If it happens that any requirement is not having a reference of the design document against it, it means that the requirement is missed out in the software design of the product. This action of updating the traceability matrix helps us in uncovering the missing requirements that have not been taken care of in the design stage. We can then modify the design to include the missing requirement.

12.5 Construction

Construction of the software product realizes the requirements. In this phase, we implement the software design. We develop the code and self-test it to ensure that the functionality as designed is achieved and is as defect-free as humanly possible. Unit testing, integration testing and system testing are part of the construction phase.

It is very common that the maximum percentage of change requests is received in this phase. Therefore, implementation of change requests is a major RM activity of this phase. Additionally, all change requests kept pending for retrofitting into the product need to be implemented during this phase.

Therefore, in this phase, we perform the activities of realizing the requirements, implementing the change requests received and updating the requirements traceability matrix to include references to the code against each of the requirements.

12.6 Testing

We carry out software testing with three objectives in mind, namely,

1. To uncover all lurking defects so as to make the product as defect-free as humanly possible.
2. To ensure that all requirements of the end users and all other stakeholders are implemented in the project.
3. The resultant software product is robust and is adhering to the design.

The three aspects, critical for the testing activity, are:

1. Test strategy.
2. Test plans.
3. Test cases.

In test strategy, we determine the best way to test the product so that the product is as thoroughly tested as possible, the cost of testing is minimized and all testing objectives are met. The test strategy is normally captured in the software quality assurance plan.

In test plans we determine the resources required to implement the test strategy, the types of tests to be carried out and the schedule of testing.

Test cases implement the test plans and draw up details of how the product is tested. A requirement normally results in multiple test cases and rarely, in only one test case.

Testing consists of designing the test cases and then executing those test cases on the software product. While designing test cases, we need to ensure that all requirements of all stakeholders are covered. Often times, it is not possible to cover all requirements in the designed test cases. The reason in most cases is either the paucity of time or funds to test the product thoroughly. Therefore, we compute a metric referred to as the "Test Coverage" metric. The formula for test coverage metric is:

TCM = (Number of requirements covered by test cases ÷ Total number of requirements as established in the requirements traceability matrix or the URS) × 100

Where TCM = Test Coverage metric.

TCM is usually expressed as a percentage. While there is no standard percentage for TCM to determine the adequacy, 90 % is usually treated as good coverage. By this statement, do not mistake me that I condone less than 100 % coverage! Far from it! Test coverage of 100 % is the best practice and I strongly advocate it.

In the testing phase, from the standpoint of requirements management, we ensure that all requirements are properly implemented in the product and update the traceability matrix with references of the test cases and test logs against each of the requirements in the matrix.

12.7 Acceptance Testing

While acceptance testing is also part of testing, it is separately handled as it is conducted by the customer. Acceptance testing is dedicated to proving that all requirements are indeed implemented in the software product. So, we need to:

1. Ensure that the acceptance test plan and test cases do cover all the requirements.
2. We need to conduct the test ensuring that all requirements are implemented in the final software product.
3. The product works without defects when used positively.

First we need to verify and validate that the acceptance test plan and test cases cover all the requirements comprehensively. Often, the acceptance testing becomes a formality before accepting the delivery of the software product. True, the acceptance testing is positive testing not intended to uncover defects but it is conducted to ensure that all end user requirements are included in the product. When the product is submitted for acceptance testing, all defects should have been uncovered and fixed. If a defect is uncovered during acceptance testing, it should be related to implementation of a requirement but not any other type of defect. But it would be wrong to assume that the earlier tests would have ensured that all requirements were implemented. Acceptance testing should focus on ensuring that all requirements are implemented and should be carried out with all diligence in positively testing the product.

Before beginning the acceptance testing, we need to verify the requirements traceability matrix to ensure that all requirements are tracked to the product and the quality records to ensure that all quality control activities are carried out diligently.

We verify the requirements traceability matrix to ensure that each of the requirements can be traced through all the software development activities and that no requirement is missed out at any stage. If any requirement is missed in the matrix, we need to lay special emphasis on testing that specific requirement.

We need to verify quality records of the project to ensure that all planned quality assurance activities are performed and that all uncovered defects are fixed. The quality records also have the potential to reveal if any requirement is missed or not implemented properly.

Thus, we ensure that all requirements are implemented properly during the acceptance testing.

12.8 Installation and Commissioning

Installation involves deploying all the machines including servers, workstations and networking equipment and then installing the software on the respective machines. Commissioning involves preparing the master data, loading it in the database, pilot runs of the system, then changeover to production and hand over the system to the users and the maintenance team.

During the installation and commissioning, especially the commissioning part of it, we will also be training the end users in the efficient usage of the system of which the application software is the most important component. In addition to training, we need to handhold them for some time so that they become adept at using the system effectively.

During both the training and the handholding period, we need to show the end users how their requirements are met by the software and how to go about performing their functions on the system and producing the results expected of them.

During this phase, we help the end users to understand how their requirements are met and how to go about achieving their results.

12.9 RM Through SDLC

Summarizing all the above discussion,

1. During the pre-project phase, we establish a need for the system so that the project can be approved and a project can be spawned.
2. During the requirements phase, we elicit, and gather requirements, analyze them and establish the requirements for the project so that software design can begin and be completed based on the established requirements.
3. During the software design phase. We ensure that all the established requirements are included in the software design so that the construction phase could realize all the requirements.
4. During the software construction phase, we realize all the established requirements in the software product and implement any and all change requests received from any of the stakeholders.
5. During the testing phase, we ensure that all requirements are indeed built into the software product and there are no defects that can be detected through the planned testing.
6. During the acceptance testing phase, we ensure that all requirements are indeed met, including the added/modified requirements to the satisfaction of the customer. The emphasis is on ensuring that all requirements are included and they are working flawlessly when used as they should be in the software product.
7. During the installation and commissioning phase, we train the end users to utilize the software product to accomplish their objectives efficiently. We handhold them to make them experts in using it to realize their stated requirements,
8. In the design phase, construction phase and testing phase, we also manage the change requests received in addition to the activities stated above.

That is how we manage the requirements through the software development life cycle.

Chapter 13
Tools and Techniques for Requirements Engineering and Management

13.1 Introduction

Requirements Engineering received significant attention from the research community as well as from practitioners. They have all focused their efforts on finding ways and means to establish software project requirements quickly, as effortlessly as possible, and as accurately as possible. Many tools and techniques were proposed for use in Requirements Engineering. The popular ones are discussed in the following sections. These are:

1. SSADM (Structured Systems Analysis and Design Method)
2. IEEE Software Engineering Standards
3. OOM (Object Oriented Methodology)
4. UML (Unified Modeling Language)
5. Agile methods

Let us discuss each of these in brief here. Each of the above is a full-fledged software development methodology from requirements to delivery. I will be giving a brief explanation about each methodology and delving deeper into how they engineer and manage project requirements.

13.2 Structured Systems Analysis and Design Method

SSADM was originally developed for the Office of Government Commerce (then it was Central Computer and Telecommunications Agency) of UK for use in procurement of software for governmental use. It has been in use since the 1980s and has been implemented in many organizations across the world, each adding its own flavor to the methodology.

M. Chemuturi, *Requirements Engineering and Management for Software Development Projects*, DOI: 10.1007/978-1-4614-5377-2_13,
© Springer Science+Business Media New York 2013

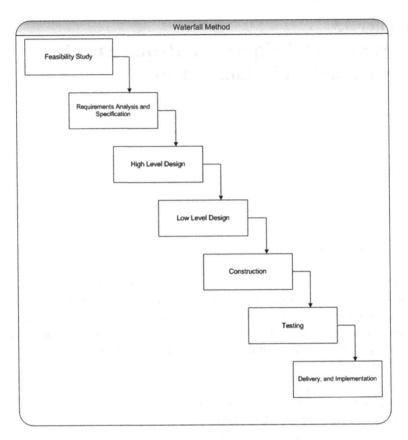

Fig. 13.1 Waterfall model

SSADM uses the waterfall method as software development life cycle. Figure 13.1 depicts the waterfall model pictorially. The waterfall model originally had five phases, namely, Requirements, Design, Implementation, Verification, and Maintenance. But since the original definition, organizations made many modifications to this model and many variants are found in the organizations. SSADM lays emphasis on rigorous documentation as the basis for the software development. This emphasis, in fact, has originated in the manufacturing model used in project type of manufacturing organizations.

Let us discuss the stages of SSADM as popularly used in many organizations, in a little greater detail.

13.2.1 Feasibility Study

Feasibility is conducted to assess the viability of computerizing a system. It collects information on the functionality being proposed, the costs, and the benefits expected to accrue from the implementation of the system. The deliverable of this phase is the feasibility report which would be considered by the management, who would either reject the proposal or accord the approval and financial sanction to the proposal.

13.2.2 Requirements Analysis and Specification

Requirements analysis and specification phase elicits and gathers requirements, analyzes them and prepares a "requirements specification" document. This aspect has been discussed in detail in the preceding chapters. Initially the requirements specification document was referred to as Functional Specifications Document and it transformed to present day's URS.

13.2.3 High Level Design

Using the requirements specification document, high level design of the system is carried out. This consisted of defining the system architecture, the main process flows (organization of the proposed system into logical modules) and defining the inputs and outputs. This is documented in a HLD (High Level Design) document.

13.2.4 Low Level Design

This phase consisted of defining the specification for each unit of software. LLD is used to carry out design of each unit of the proposed software product and the specifications are then documented. The deliverable from this phase is the LLD (Low Level Design) document. The software programming would be carried out using this LLD document.

13.2.5 Construction

During this phase, all programs would be constructed and subjected to unit testing. The completed units are then integrated into their modules and the modules are subjected to integration testing. All defects uncovered during the tests are fixed. The deliverable of this phase is the working source code and executable code.

13.2.6 Testing

In this phase, the executable code is subjected to system testing on the target system on which the software is to be implemented and any defects uncovered are fixed. The deliverable of this phase is the readiness of the system to roll into piloting, parallel runs and production.

13.2.7 Delivery and Implementation

During this phase, the software is delivered to the target system and is implemented, including conducting pilot runs of the system, running it in parallel with the existing system and then finally rolling out the new system into production. This phase also includes the preparation of user documentation including, the user manual, operations manual and troubleshooting manual.

13.2.8 Software Maintenance

The next activity on software is obviously the maintenance of the software in production. Maintenance includes bug fixing, modification and functional expansion. The software would be replaced when a major change to the system becomes necessary from the functionality point of view or due to a major technological change such as obsolescence of the existing hardware or the onset of a phenomenon like the Y2K or the Internet.

SSADM also devised a few tools for modeling and documenting the requirements and design. These are:

1. **Logical Data Modeling**—It is the representation of a system which can be either a manual system or a computerized system. During requirements analysis, we build a logical data model of the manual system being taken up. During design, we build a logical data model for the proposed computerized system. Logical data model was proposed during the days of Hierarchical DBMS (Database Management) and Network DBMS. With the onset of Relational DBMS, ER (Entity Relationship) modeling was developed from the logical data modeling.
2. **Dataflow Modeling**—It is building a model of the flow of data in the system. As computerized systems basically process data, the system processes revolve

around the data. Each process transforms a part of the system data in some way. When all of the processes are executed, the system data is transformed as desired. The data flows from one process to another until information is extracted and presented as output of the system. This model is presently referred to as process modeling. During requirements analysis, we model the existing processes and during design, we model the proposed system.

3. **Entity Behavior Modeling**—It is modeling the sequence of operations in the system. Events in the processes impact and transform the entities. We model each event and the behavior of the impacted entities and the sequence of events in a process from the first input to the final output, diagrammatically in this modeling.

SSADM, from the day it was implemented, held its place among the other various software development methodologies. The original model was modified by many organizations and researchers. I would rather go to the extent of saying that all other methodologies that have emerged since, still contain a streak of SSADM in them. In the spiral model and the iterative model, each increment/iteration is still a waterfall model. The logical data modeling which later transformed into ERD (Entity Relationship Diagram) is the only way to model system data even today. The process standards like the ISO 9000 and CMMI® implement the philosophy of SSADM. IEEE also followed the SSADM model in defining the software engineering standards when they released the first set in 1988. The agile methods denounce the heavy reliance of SSADM on documentation but still implement the waterfall model in each of the iterations. In outsourced development SSADM is still significantly utilized to drive software development contracts.

13.2.9 Requirements Engineering and Management in SSADM

SSADM accorded significant importance to the aspect of requirements engineering and management by designating a separate phase for this activity. During this phase, the activities of elicitation, gathering, analysis and establishment are carried out. The deliverable of this phase is the Requirements Specifications Document.

The technique of logical data modeling to model the system data was the originally suggested technique but now, ERDs are being used to model data. The technique of dataflow diagrams is used to model process flow in SSADM.

13.2.10 ER Diagrams

Data modeling deals with establishing the relationship between various data entities in the system. In any information transaction, data is transmitted between entities. Let us take a simplified purchase transaction:

1. An item is to be purchased in an organization. So a purchase order is raised on a vendor.
2. The vendor supplies the item to the inventory.
3. The vendor raises an invoice on the organization for payment.
4. The organization makes the payment.

From this transaction we could establish relationships:

1. There is a relationship between the purchase order and the vendor
2. There is a relationship between the vendor and the inventory
3. There is a relationship between the vendor and the invoice
4. There is a relationship between the invoice and the payments

Modeling relationships of this type is referred to as data modeling.

An entity is a place, person or a thing and is described by its attributes. For example, an employee in a payroll system is an entity. A purchase order is an entity in a material management system. ER Diagrams (ERD) pictorially represent the relationships between various data entities in the system.

The entity is represented by a rectangular box in the ERDs. The relationship between entities is represented by a line in ERDs. The ends of the line represent the type of relationship between entities. There are three types of relationship:

1. One-to-one relationship—An item would be in one purchase order. This is represented in the ERDs by a straight line with normal end.
2. One-to-many relationship—A purchase order could contain many items. This is represented in the ERDs by a straight line with crow's feet at the end of the line having "many" relationship.
3. Many-to-many relationship—A vendor can receive multiple purchase orders and supply multiple materials. This is represented in the ERDs by a straight line with crow's feet at both ends of the line.

The symbols used in ERDs and the relationships are shown in Fig. 13.2.

Figure 13.3 depicts a simple ERD, modeling the purchasing transaction detailed above. A purchase order entity is placed on the vendor and the vendor entity may receive multiple purchase orders as depicted in the figure. The vendor would supply items to warehouse inventory. The vendor could be supplying multiple items to the warehouse against multiple purchase orders. The vendor would raise invoices for materials supplied. The warehouse would be requesting purchase of items which have fallen below the reordering level set for the items.

Of course, real life ERDs would be much more elaborate and complex as there would be many entities in a system and they would have complex relationships with each other. The real life ERDs would span across many sheets. Many software tools are available for modeling entity relationships and some of them could be using different types of notations, especially in representing the relationships between entities.

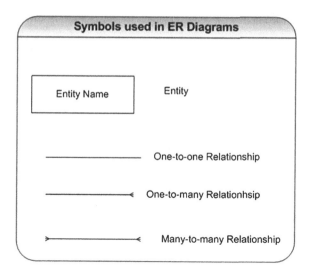

Fig. 13.2 Symbols used in ER diagrams

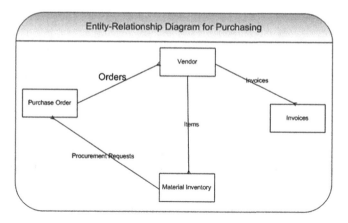

Fig. 13.3 Example of an entity-relationship diagram

13.2.11 Data Flow Diagrams

Dataflow diagrams are used to model the process of transforming the data by the system. During the requirements phase, we model the existing system and during design phase, we model the proposed system.

The symbols used in DFDs are shown in Fig. 13.4. The symbol of the process has many variants. I am depicting two of those variants.

Fig. 13.4 Symbols used in
DFDs

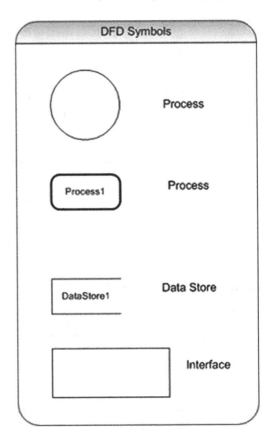

Now let us model the purchase order transaction using a DFD. Let us first enumerate the steps in the process which we can model pictorially. Here is the simplified process:

1. An executive from the production department or warehouse raises a procurement requisition on to the purchase department for purchase of item or items.
2. The purchase department receives the requisition and files it (stores it).
3. The purchase department raises enquiries on vendors asking for price quotations.
4. The vendors receive the enquiry and stores it.
5. The vendors transmit price quotes to the purchase department.
6. The purchase department receives the quotes and stores them.
7. The purchase department transmits the quotes to the executive who originated the procurement requisition.
8. The executive selects the vendor and sends the recommendation to the purchase department.
9. The purchase department stores the recommendation.

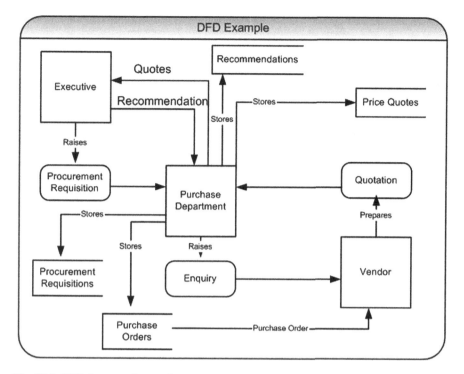

Fig. 13.5   DFD for a purchase order process

10.   The purchase department raises a purchase order on the selected vendor.

Let us model this process in a DFD. Figure 13.5 depicts this process pictorially.

DFDs for real life systems would span across multiple sheets. To make a large system comprehensible, it is normally divided into multiple levels. The top level DFD is normally a context diagram described in the subsequent sections. Then for each subsequent level a DFD is prepared. The lowest level DFD would be for a software unit. If a totally granular DFD set is prepared, it can almost supplement a design document. But more often than not, DFDs would not be prepared to the lowest unit level. Each organization decides at what level of granularity, the DFDs would be stopped in its projects.

13.2.12   Context Diagram

Context diagrams are used to show the context in which the proposed system operates. It also shows the context of the modules within the system. In the context diagrams, circles are used to represent entities and arrows to show the relationship with arrows pointing in the direction of the flow of information between the entities. An arrow head only at one of the ends depicts a unidirectional

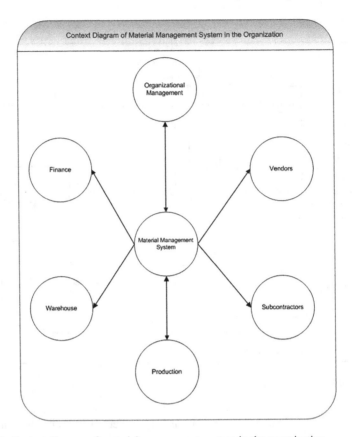

Fig. 13.6 Context diagram of material management system in the organization

relationship. That is information flows only in one direction. A line with arrow heads at both the ends depicts a bidirectional relationship. That is, information flows in both directions between the entities.

Figure 13.6 depicts the context of material management system in an organization.

13.2.13 Structure Chart

S structure chart is used to depict the hierarchy of the functionality in the system. It uses rectangular boxes to depict entities in the system and the lines with arrow heads show the flow of information between the entities. Figure 13.7 depicts a structure chart pictorially.

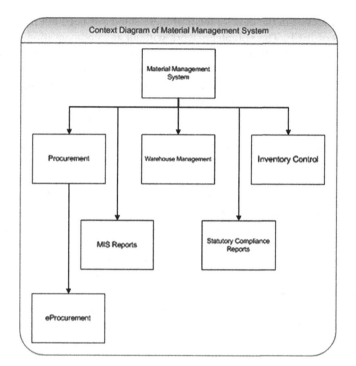

Fig. 13.7 Structure chart of material management system

There could be many variants of the model and the diagrams described above. SSADM is one of the most researched models of software engineering methodologies. Many researchers modified the SSADM with their own improvements and authored books about their models. Organizations too have customized the model and implemented it in their own way. It is not an exaggeration to say that SSADM has not been implemented in its original form anywhere except in the beginning and that too in UK. So, I do not claim that the model I described here is the "right" SSADM but what I described is the one I have observed in many organizations.

13.3 IEEE Software Engineering Standards

IEEE (Institute of Electrical and Electronics Engineers, USA) is an association of engineers from all over the world. Its membership, of about 400,000, includes engineers with a minimum of graduate level qualifications in the electrical, electronics, computers and telecommunications engineering or engineers working in these fields. IEEE has a standards wing which develops standards for the industry collaboratively with engineers drawn from the industry and the academia. The individuals involved in the development of these standards need not necessarily be the members of IEEE. The volunteers work without any compensation too. All the standards developed by IEEE are subject to peer review and voting before they are released. All the standards are periodically reviewed and the revised versions are regularly released. IEEE standards are highly respected and are implemented in the products and interfaces all over the world.

I am proud and feel privileged to be admitted to the IEEE *in the first place and be elevated to the category of Senior Member in the second place.*

IEEE undertook development of standards for the field of software engineering and released the first set in 1988. Some of these are revised and re-released in 1997–1998. There is great wisdom in these standards. Many organizations have adopted these standards. The process model CMM® (Capability Maturity Model) released by SEI (Software Engineering Institute of the Carnegie Mellon University) in 1998 emphasized IEEE standards in its model document, even though, this emphasis is dropped in its later versions of CMMI® (CMM Integration).

IEEE standards advocate a methodical and process driven approach. One important aspect to be noted here is that implementing IEEE standards has not failed any organization so far!

Table 13.1 gives a list of software engineering standards released so far.

All of these standards promote methodical working and implementation of industry best practices. They focus on large scale projects but allow scaling to suit smaller projects. The standards relevant to Requirements Engineering and Management are, 610, 830, 1028, 1028, 1044, 1233 and P1805. IEEE continues to develop and improve the standards and some more standards will continue to be released. IEEE has initiated a project (P1805) on the language to be used for defining requirements for software projects. It may be released soon.

Implementing IEEE standards in the organization is one of the best practices, in software development in general and REM in particular.

13.4 Object Oriented Methodology

Object Oriented programming brought real-world thinking into software development. In the real-world there are objects that have characteristics and functions. How the objects perform and use the characteristics and produce results is not of concern to the outside world. Therefore, the programs were modified to resemble real-world objects.

Object oriented methodology is a software development methodology that views software development as development of objects (instead of programs) that can be assembled into a software product. Each object is not a complete stand-alone unit but is a component that can be picked up and used along with other objects to assemble a software product. Each object has data structures built into it along with the methods (small programs or functions or subprograms) that utilize it. The object encapsulates (conceals) the methods and the data structure from

Table 13.1 List of software engineering standards

Standard number	Brief description of the standards
610	IEEE standard glossary of software engineering terminology
730	IEEE guide for software quality assurance plans
828	IEEE standard for software configuration management plans
829	IEEE standard for software test documentation
830	IEEE recommended practice for software requirements specifications
982	IEEE guide for the use of IEEE standard dictionary of measures to produce reliable software
1008	IEEE standard for software unit testing
1012	IEEE standard for software verification and validation
1016	IEEE recommended practice for software design descriptions
1028	IEEE standard for software reviews
1044	IEEE guide to classification for software anomalies
1045	IEEE standard for software productivity metrics
1058	IEEE standard for software project management plans
1061	IEEE standard for a software quality metrics methodology
1062	IEEE recommended practice for software acquisition
1063	ieee standard for software user documentation
1074	IEEE standard for developing software life cycle processes
1175	IEEE trial-use standard reference model for computing system tool interconnections
1219	IEEE standard for software maintenance
1220	IEEE standard for application and management of the systems engineering process
1228	IEEE standard for software safety plans
1233	IEEE guide for developing system requirements specifications
1320	IEEE standard for conceptual modeling language and syntax and semantics for IDEF
1348	IEEE recommended practice for the adoption of computer-aided software engineering (case) tools
1362	IEEE guide for information technology—system definition—concept of operations (conops) document
1420	IEEE standard for information technology—software reuse—data model for reuse library interoperability—basic interoperability data model
1430	IEEE guide for information technology—software reuse—concept of operation for interoperating reuse libraries
1471	IEEE recommended practice for architectural description of software-intensive systems
1517	IEEE standard for information technology—software life cycle processes—reuse process
12119	IEEE application of international standard ISO/IEC 12119—information technology—software packages—quality requirements and testing
12207	IEEE/EIA guide—industry implementation of international standard ISO/IEC 12207—standard for information technology—software life cycle processes—implementation considerations
14143	Implementation note for IEEE adoption of ISO/IEC 14143 information technology—software Measurement—functional size measurement
P1805	Guide for requirements capture language (to be released)

outside view. That is, the user need not be concerned with the "how" of the object. The user can view the object as a black box that receives some needed inputs and outputs the expected values. Since each object is self-contained, they can be easily reused. Thus object oriented methodology avoids the need for re-programming for the same functionality.

The following terminology is associated with OO Methodology. This is given to introduce the reader to the concepts underlying the OOM. This explanation is given in brief and the reader is advised to refer to other material for full coverage of the subject, if felt necessary.

Object—an object is a combination of methods (functions/subprograms/small programs) and data structures that are used by the methods. Each of the methods performs one action and achieves a predefined functionality affecting the data defined in the data structures.

Class—a class is a model of the real world from which an object can be created. A Class is abstract and object is its implementation. A Class is a "super object" in that every object is an instance of a class.

Message—A Message is the input to the object that invokes a method contained inside the object and spurs that method into action to process and produce a response to the message received. This response is again conveyed as a message back to the originator.

Abstraction—is the action of analyzing the real world objects and forming classes based on the similarity of characteristics of the real world objects so that they can be understood, analyzed, designed and implemented to produce the desired software product. Abstraction is at a high level. It does not consider the implementation details.

Encapsulation—In object oriented methodology, the data is hidden from the sight of the users. The data is accessed through the methods contained in the object. This aspect of preventing direct access to the data of the objects is referred to as encapsulation.

Polymorphism—it is the ability of the objects to be implemented differently to achieve multiple functionalities. When the objects are similar but perform differently, the same message can be used to communicate with different objects and obtain different results. The ability to obtain different results using the same message by sending it to different object is referred to as polymorphism.

Inheritance—Inheritance is the ability of the object to inherit the characteristics of the class from which it is instantiated. When we instantiate an object from a class, the object inherits all the methods and data structures of the class. In addition, the object can have some more methods and data structures.

OOM focuses more on the engineering side of software development than on the management side. Therefore, it does not tell us how to go about managing the software project using OOM. One thing is clear though, OOM expects that the user requirements are already established. It is a pre-project activity as far as OOM is concerned. OOM starts with analyzing the user requirements and then extracts the classes from the requirements using abstraction, then designs classes, and implements them in the software code. Then the software product is assembled using the objects instantiated from the classes. The product is then tested and deployed. This is in brief how the object oriented software development project is executed.

So, when we come to the requirements management portion of the project, we need to establish the user requirements as we would in any other software project. OMT (Object Modeling Technique) was oriented towards modeling the design but not the requirements and therefore, it is not covered here. OOM is currently using UML, which is detailed in the following sections for modeling the design as well as requirements.

13.5 Unified Modeling Language

Unified Modeling Language (UML) is used for modeling requirements and design of software systems. It began in the object oriented methodology but is currently used in all types of software development projects. UML was created and is maintained by the OMG (Object Management Group). Ivar Jacobson, James Rumbaugh, and Grady Booch are credited to have created UML at Rational Software (now a part of IBM).

One aspect to remember is that UML is not a software development methodology but is meant to model the computer applications. It is more like a language to describe the system. UML uses the following diagrams to model the software system:

1. Class Diagrams
2. Use Cases
3. Sequence Diagrams
4. Statecharts
5. Activity Diagrams
6. Component Diagrams
7. Deployment Diagrams

We will discuss each of them in a little greater detail in the below sections.

13.5.1 Class Diagrams

Class diagrams are used to depict the classes in a model. Class diagrams are used to model high level design (roughly equivalent to SRS). User requirements are analyzed and classes are abstracted and then modeled using class diagrams. Each class has attributes (data), methods, and relationships with other classes. Figure 13.8 depicts the symbols used in class diagrams. Using these symbols, class diagrams are prepared to model classes in the system. Figure 13.9 depicts a very simplified class diagram for a procurement system.

Fig. 13.8 Symbols used in
class diagrams

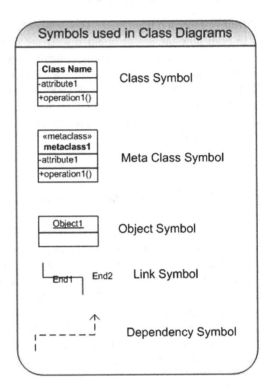

13.5.2 *Use Cases*

A use case (a case in the use of the system) depicts pictorially a "unit of functionality" of the proposed system. Usually, a use case has two elements, namely, the use case diagram and a use case description.

Use case diagrams use the symbols depicted in Fig. 13.10. In use case diagrams, an ellipse represents the use case. It is usually accompanied by the name of the use case and optionally a use case ID. The actor is depicted by the symbol of a stickman. The stickman is usually identified by a name. The actor could be a human being interacting with the system using a GUI (Graphical User Interface) or another system interacting with the system using a machine interface or a protocol. The interaction between the actors and the use case is represented by lines. The system boundary is represented by a rectangle. The actor is usually outside the system. Figure 13.11 depicts the procurements system using use case methodology.

The description that accompanies a use case diagram can be a free flowing scenario description but a structured description is preferable. In whatever form the organization desires to document the description, it is better to include the following options in the use case description:

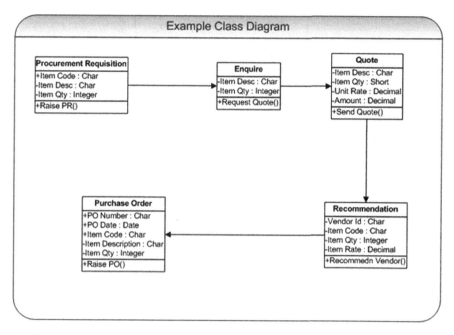

Fig. 13.9 Example of a simple class diagram

Fig. 13.10 Symbols used in use case diagrams

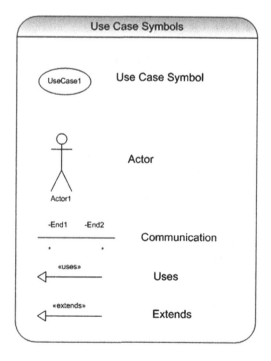

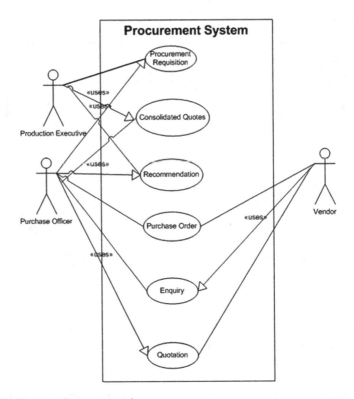

Fig. 13.11 Use case diagram example

1. Use case name and optionally use case Id
2. Primary actor
3. Other actors
4. Objective of the use case
5. Precondition
6. Trigger—the event that triggers this use case
7. Exit condition (the definition of completion or success of the use case)
8. Use case description
9. Workflow
10. Alternate workflows

An example description of the use case depicted in Fig. 13.11 is given in Table 13.2.

In this manner, we describe each of the use cases in the use case diagram. Now, as you can perhaps see, some vital information is missing from the above description. The data description is more or less absent. For each of the data items, the details of data type (numeric, character, date, currency etc.), the field width, their constraints etc. are not mentioned. In some cases, these are described in the use case description or a separate entry is used to describe this information. In some

Table 13.2 Description of the use case (Procurement Requisition) depicted in Fig. 13.11

Project Id	
Use case name	Procurement requisition
Primary actor	Production executive
Other actors	Purchase officer, uses this use case to raise enquiries of prospective vendors
Objective of the use case	1. To capture the material requirement of the production executive
	2. To obtain full information of the items to facilitate raising enquiries
Precondition	There should be a production order against which the expenditure can be booked
Trigger	The BOM (Bill of Material) should have been prepared and approved
Exit condition	1. Procurement requisition is filled in by the production executive
	2. Email is sent to purchase officer that a procurement requisition is awaiting for next step
Use Case description	After a purchase order is received from a customer by the marketing department, it will be passed on to the engineering department. The engineering department would prepare engineering drawings for manufacturing the product and releases them to the production planning. Production planning raises production orders to various production shops authorizing them to initiate production. Production shops assess the material stock and raise procurement requisitions on purchase for procurement of materials that are in short supply. This requisition triggers procurement action. This requisition would contain information about the item code, item description, the required quantity, the date by which the item is required and the estimated cost of the item being procured, along with references to the production order and the project id. The requisition is computer based and need to be accessible from the production shop's PC
Workflow	1. Production order is received by the production shop
	2. Material stock in the warehouse is assessed and material shortages are enumerated by the production executive
	3. Production executive raises the procurement requisition using the computerized procurement system
	4. The requisition is sent for authorization of concerned authority automatically by the computer
	5. The authorization is granted by the concerned executive
	6. An email is sent to the purchase office that a procurement requisition is awaiting action
Alternate workflow	Sometimes, the system itself has to raise the procurement requisition especially in the matter of standard items such as hardware like nuts and bolts, stationary, cleaning materials and so on. This has to be based on re-order level, ordering quantity, and safety stock decided for each of the standard items in the stock

cases, they attach formats currently being used in the organization. Sometimes, these are elicited from the users by the designers and are used for software design.

Use cases have become very popular tools for capturing the project requirements in recent times due to their simplicity and clarity of presentation. The use case diagrams are very easy to draw and even easier to comprehend.

The data description is absent and it needs to be obtained and documented using other means. The use case diagrams are indeed simple but it is difficult to represent complicated logic in large complex software system.

In the Table 13.2, I have not included the description for all the use cases included in the Fig. 13.11. Only one use case is described as an example. Using the format given in Table 13.2 other use cases can be described.

Use case diagrams coupled with use case descriptions are used to capture requirements for the projects by many organizations.

13.5.3 Sequence Diagrams

Sequence diagrams depict the sequence of operations between classes or objects. Sequence diagrams are used more to model the design but can also be used to model requirements. Sequence diagrams depict the classes/objects included in the scenario and the interactions between them along with the sequence.

Objects are depicted using rectangles, the actors are depicted using the stickman, and the messages are depicted using solid lines and dotted lines. Figure 13.12 depicts a sample sequence diagram for the procurement system.

13.5.4 Statecharts

Statecharts are used to model the behavior of the entities in the system. Usually, statecharts are more often used to model the design of the system but they could also be used to model the behavior of the use cases

A very simple statechart is depicted in Fig. 13.13. In this statechart, the requirement of material is recognized and when the event of Production Executive raising the procurement request the state of material requirement transitions to Procurement Requisition. Now the event of the Purchase Officer raising an enquiry on the prospective vendors transitions the state to Enquiry.

Statecharts are used more to depict the design of the proposed system than in capturing the requirements. All, the same, statecharts are definitely useful in capturing the requirements especially the transition of the state of the entities.

13.5.5 Activity Diagrams

Activity diagrams model the procedural flow of actions that are part of a use case or a set of use cases. Activity diagrams are normally used to depict the sequence of execution in a system. Activity diagrams use similar notation as statecharts.

A sample activity diagram is depicted in Fig. 13.14 for the procurement system in a simplified manner.

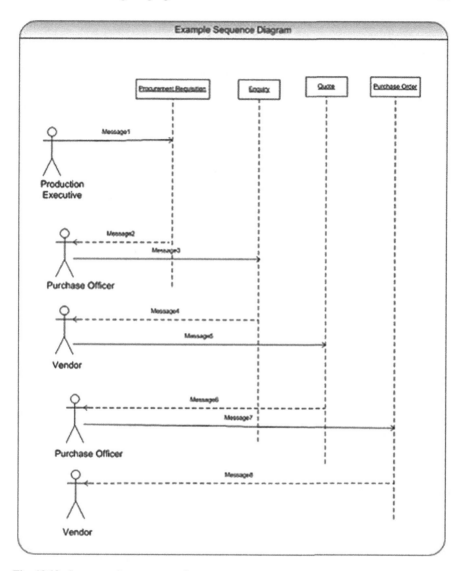

Fig. 13.12 Sequence diagram example

13.5.6 Component Diagrams

Component diagrams model the relationship between the components of a software system. Component diagrams are used to model the design of a software product rather than to capture the requirements of a proposed project. I am not going into the details of this type of diagramming as it is not used for capturing

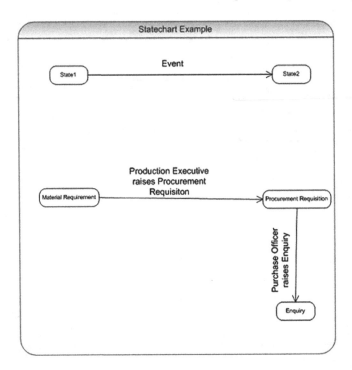

Fig. 13.13 Example of a simple statechart

requirements and this book's focus is on requirements engineering and management.

13.5.7 Deployment Diagrams

Deployment diagrams are used to depict the physical deployment of system artifacts including the hardware artifacts, system software artifacts and application artifacts of a facility. They deal with the facility management and system design that includes hardware as well rather than with either project requirements of software design. Deployment diagrams are not used in the field of requirements engineering and management. Therefore, I am not covering them in this chapter.

13.5.8 Final Words on UML

UML has become very popular of which use cases have been particularly popular, because of their simplicity and ease of use. Use cases along with use case descriptions

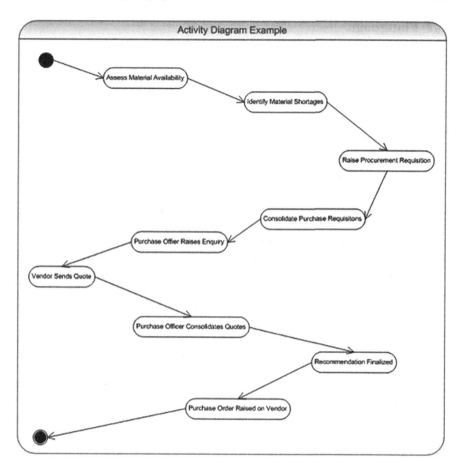

Fig. 13.14 Activity diagram example

are being used to describe project requirements. Of the other diagrams, class diagrams and sequence diagrams are also used extensively. Others are being used but not on the same scale as the use cases.

13.6 Agile Methods

Agile methods do not use any formal methods for modeling the systems either for capturing the project requirements or the proposed system design. If at all, they use UML modeling techniques described above for either capturing project requirements or the design of the proposed system. We have a more detailed discussion about agile methods in Chap. 12.

13.7 Planguage

Planguage is developed by Tom Gilb for describing requirements as well as software design. It is dealt with in much greater detail in Appendix B and therefore not described here.

13.8 Final Words on Tools and Techniques in Requirements Engineering and Management

All the tools and techniques described above, more or less, do aid requirements engineering rather than management. Requirements management is basically ensuring that the activity is carried out efficiently and effectively. Project plans ensure that the activity is planned as described in Chap. 7 on planning. Ensuring that requirements are included and fulfilled at every stage of software engineering is achieved using the requirements traceability matrix which is discussed in Chap. 9. Changes are inevitable midway through any human endeavor and requirements management is no exception. We discussed requirements change management in Chap. 8. The important aspect to be defined for effectively carrying out the activity is the assignment of responsibilities to appropriate agencies and sharing it between the organization and the individual. This aspect of roles and responsibilities is discussed in Chap. 11. The tools and techniques useful in these activities are discussed in the cited chapters.

Now, let us address the question—which tool or technique is best suited to a given scenario? I confess that there is no single right answer to this question. All the tools and techniques discussed in this chapter help us in modeling the system and capture the requirements so that we can understand it fully and thereby design and build the "right" software system for our clients. Another important point I would like to stress here is that the techniques described above are by no means comprehensive. There are a plethora of models and diagrams available in software engineering. Just to record all the models and diagrams available would take a full book in itself.

One tool that is used across the board for modeling data and assisting us in the design of the database is the ER diagram. I have not seen any other technique being used to model data as much as ER diagrams are used.

But when it comes to modeling project requirements, there is diversity. SSADM is still very much in use and looks to be there for some more time to come. UML has stormed on to the scene and has been put to use in significant number of projects. But it has been criticized too. In fact Ivar Jacobson who perhaps is more responsible than anyone else in defining the UML has started another initiative to define a new methodology for software engineering. The web site www.semat.org is spearheading that initiative. Semat stands for Software Engineering Method and Theory. It is proposed to release the new methodology by the year 2013.

What I can certainly say with confidence is that the software engineering tools are nowhere near the granularity or the accuracy that other engineering disciplines have in engineering drawings. They say that "drawing is the language of engineers." But it is not so for software engineers. We, the academics, the practitioners, and the thinkers, in the software engineering field seem to continue our search to find the perfect set of tools so that we can define our requirements and design without ambiguity and in a manner that all concerned would interpret it in the same manner.

In the meanwhile, we continue to use the tools and techniques with which we are most knowledgeable and comfortable to use.

Chapter 14
Pitfalls and Best Practices in Requirements Engineering and Management

14.1 Introduction

Requirements engineering and management is a vital activity in software development. If other activities are not handled properly, there could be defects in the product or the project may suffer from cost or schedule overruns. But if requirements are not handled properly, we may have a wrong product on our hands at the end of the project! Requirements management has a strategic impact on the project. It is the difference between success and failure of the project. Defects can be fixed, but we cannot re-develop the product.

Therefore, it is imperative that we implement the industry best practices and avoid the pitfalls in the management of requirements if we wish to ensure success of the project by delivering the right product. Here we discuss some of the pitfalls that frequently plague the requirements management activity as well as the best practices that have always ensured the success of the project.

14.2 Best Practices and Pitfalls at the Organizational Level

Organization has a major role in either the success or failure of the project. Organization sets the framework in which the individuals perform and can excel or fail. The framework facilitates the individual capacity to perform. It can propel the individual to excel or push the person to fail. Organization has the onus in the following aspects:

1. Approach to Requirements Engineering and Management
2. Provision of Resources
3. Training and Updating of Skills
4. Definition and Improvement of Process
5. Motivation and Morale of the Resources

M. Chemuturi, *Requirements Engineering and Management for Software Development Projects*, DOI: 10.1007/978-1-4614-5377-2_14,
© Springer Science+Business Media New York 2013

6. Quality Assurance
7. Knowledge Repository.

Let us discuss each of the above aspects in detail hereunder.

14.2.1 Approach to Requirements Engineering and Management

This is perhaps the first and most important aspect of the organizational framework the organization has the responsibility to set. There are basically two approaches, namely,

1. A process driven approach
2. Ad hoc approach.

A process driven approach consists of defining a process that is appropriate for the organization that includes industry best practices as well. All of the activities are carried out in the organization conforming to the corresponding process. When, it is essential, a waiver is requested and taken from the concerned authority to perform an activity deviating from the specification of the process. A waiver is usually granted only when the suggested approach is argued to be better than the existing one. A process driven approach ensures predictable results. It makes a novice perform like an experienced one and an experienced one to perform like an expert. A process driven approach also brings in uniformity across the organization and the customer can expect similar results irrespective of the person performing the activity. There is a criticism that a process driven approach curtails the freedom of the individuals performing the activity. This is partially true but waivers are granted in deserving cases although unbridled experimentation is not permitted.

An ad hoc approach gives all the freedom to the individuals performing the work. There will be no process to conform to. In these cases, the results depend on the expertise, experience and the motivation of the person. If the person is an expert, has excellent experience and is highly motivated, this approach can produce spectacular success. But if the person is a novice and is poorly motivated, the results can be spectacularly disastrous.

For small organizations running fewer than three projects concurrently, an ad hoc approach may be adequate as the gaps in knowledge can be bridged by the superiors. But as the organization grows larger and handles many projects concurrently, an ad hoc approach may not be adequate.

When organizations start, they are usually small handling few projects concurrently. Then they grow larger with their initial success. Then if the growing organization does not then adopt a process driven approach, it may not be able to control the project execution efficiently and may move toward failure. So, normally organizations move towards a process driven approach as they grow in size.

The best practice is to adopt a process driven approach in the early stages of the organization itself so that the projects are executed in a disciplined manner from the beginning.

14.2.2 Provision of Resources

Organization provides resources without which even work itself cannot be performed. But often organizations overlook on the quality of resources. Some organizations entrust the requirements engineering activity to software engineers themselves who are not knowledgeable in the project's domain. In some cases, senior software engineers are assigned and in some cases, even programmers are assigned to the requirements activity. When the resources are not properly qualified or trained, and are assigned to perform an activity, the activity suffers on both the quality and productivity fronts. Obviously, the requirements are not defined properly resulting in many change requests during the project execution. Too many change requests affect the rhythm of the software development resulting in cost and schedule overruns at a minimum.

The best practice in this area is to provide qualified resources, with the right kind of training in the requirements engineering activity, tools and techniques thereof, as well as, mentoring on the job to make them full-fledged business analysts. Only such people ought to be assigned the activity of requirements engineering. When it becomes necessary to assign software engineers to requirements engineering activity for any reason, it is essential to provide them training in requirements engineering and the project domain at a minimum before assigning them the requirements engineering activity.

The Second pitfall that I have frequently observed is that inadequate time is allocated to requirements engineering. Managements often push the project team to begin coding before the project requirements are properly and comprehensively established. Why organizations fall into this pit is because the project is allocated with all the resources—all at once. It is better to space out the allocation of resources on an "as required" basis. That way, the pressure to utilize the programming resources would not exist and allows time to perform the requirements engineering activity comprehensively. The best practice is to allocate an adequate duration for carrying out the requirements engineering activity. I am not advocating spending unending time in carrying out the activity but I do advocate giving the requirements engineering activity due consideration and importance while preparing the project schedule. The right amount of time needed for a project depends on the nature of the project and the amount of work involved in establishing the requirements. What we need to do is to give this activity due importance and consideration during project estimation and scheduling so that the activity is performed diligently and comprehensively.

14.2.3 Training and Updating of Skills

Training has been the bane of many software development organizations, Training of resources is not considered as an important activity. Training imparts skills in the human resources necessary to carry out the present job effectively, as well as, to equip them to be ready to shoulder higher responsibilities in the future. The right kind of training goes a long way in ensuring that the performance on the job is effective. While organizations spend readily for imparting training on a new language programming, they do shy away from spending on imparting requirements engineering skills to selected employees.

The software development field is one noted for fast obsolescence through continuous new developments. The requirements engineering itself has undergone a metamorphosis from the days of a simple functional specifications definition to present day's advanced requirements establishment techniques. Therefore, the skills of the people carrying out the requirements engineering ought to be upgraded on a regular basis. This can be achieved by sponsoring employees to attend public seminars conducted on the topic, subscribing to related journals, conducting knowledge sharing sessions and so on.

The pitfall has been that these activities of training and updating of skills had not been given adequate attention in the organizations. The best practice is to conduct initial skills training and then periodically update the skills with the latest developments in the field using a process driven approach.

14.2.4 Definition and Improvement of Process

Any organizational activity would have predictable results if it is driven by an organizational process which is continuously improved in line with the changing times. It is the organizational responsibility to define a process for carrying out the requirements engineering activity and to continuously improve it.

The pitfall has been that organizations do not define a robust process, including procedures, standards, guidelines, formats, templates and checklists to carry out requirements engineering activity. Sometimes a sketchy process would be defined under the garb of allowing freedom to the people performing the activity. Sometimes a process is defined but it is relegated to the records and not implemented. Sometimes the process is not improved on a periodic basis. The ills plaguing the industry in terms of the process definition and improvement are many. This has been the pitfall of many organizations.

The best practice is to initially define a robust process appropriate for the organization and to improve it continuously using a process driven approach.

14.2.5 Motivation and Morale of the Resources

Provision of resources, training them and updating their skills, as well as, definition and improvement of a robust process would all be brought to naught if the individuals performing the activity are not motivated and their morale is not maintained at a high level. A highly motivated set of individuals can achieve miracles and a badly motivated team would not be able to achieve even moderate success. We see this happening on the sports field regularly and history is replete with many examples.

How do we motivate the individuals? It is not easy to answer this question in simplistic terms. The need for motivation differs from individual to individual. What motivates one individual may not motivate another. Money and fear, the traditional tools used for motivating individuals have lost their sheen. The implementation of need-based-minimum-wage concept and fair employment practices have been done away with the capability of fear to motivate. They also dented the capacity of money to motivate. Some individuals are still motivated by these two tools but only for a limited duration after which they lose fear. Money to some extent is still a motivator but it is not getting corresponding return on investment.

There are many theories of motivation and I am referring to only two of them. The first one is the "carrot-and-stick" theory and the other is the "expectancy" theory. These are in my opinion, are very important to apply at individual level.

When people come to work in an organization they do have some expectations, such as payment of wages regularly, fair treatment, possibility for advancement in career, rewards for performance beyond the normal level, punishment for bad performance and so on. When these expectations are met, the individual stays motivated to perform. When any of these expectations are not met, the motivation of the individual deteriorates. So it delves upon the organizations to meet these expectations of the employees. It would be very difficult for the organization or the individuals to meet these expectations, if:

1. These expectations are not recognized at all
2. The policies concerning good performance, and bad performance are not defined or not transparent to the individuals
3. The policies regarding career advancement are not defined or not transparent to individuals
4. The policies regarding reward and punishment are not defined or not transparent to the individuals.

So, it is imperative for the organizations to define these policies, make them transparent to the employees and implement them scrupulously so that the expectations of the employees can be met satisfactorily. Defining these policies and making them transparent would set the right expectations in the employees. Often organizations do promote employees to the next level, provide rewards and recognition, and discipline erring employees without any explicit definition.

The activities are performed, perhaps on a regular basis, as I saw in many organizations, but these are not transparent to the employees. These are not based on any policy definitions, or set targets. When this happens, employees suspect cloak-and-dagger methods and this sets wrong expectations in the employees. Wrong expectations of employees can never be met leading to demotivation of the individuals. The best practice is to set right expectations in the employees by explicit declaration of relevant policies.

The carrot-and-stick theory implies reward for good work and punishment for bad work. When this theory is implemented scrupulously it would yield wonderful results. But any laxity in its implementation or any instances of non-implementation would water down the results. As Douglas McGregor said the discipline should be like a hot stove. It should cause a burn without any bias to anyone who touches it; the burn should be commensurate with the amount of touch; and the burn should be immediate. If handing the stick is not handled in the way of a hot stove, it is bound to produce unpredictable results. The pitfall of many organizations is to deviate from the hot stove theory when handling disciplinary cases.

There is a misconception that all the rewards ought to be financial in nature. They need not be. Many people are motivated by recognition more than by money. Most people crave more for affection and recognition than money. A meeting with the Chairman of the company motivates a person much more than a week's salary, perhaps. Many organizations do give rewards to their employees, some financial and some non-financial in nature. While they do so, it sometimes happens that the same person keeps getting the reward successively every time. The person may be a super performer and richly deserves it having earned it by sheer performance, every time. But this would have a demotivating effect on the others who just stop trying for the award being unable to compete with the super performer. When we have a super performer on our hands, we ought to find different ways of motivating that individual. We may promote him/her or give a higher rate of pay. But we need to give the reward to others too. All the employees should see the possibility of winning the award. Only then, they would try to excel.

Similarly, modern organizations are shying away from giving negative rewards (punishments). In the days gone by, organizations relied entirely on negative rewards and the world has come full circle with the organizations relying only on positive rewards. While positive rewards propel employees towards better performance and excellence, negative rewards keep them from indiscipline and work to the detriment of the team performance. Negative rewards are the ones that prevent a person from being selfish and goad the individual towards cooperating with the team. Ultimately organizations require success of the team. Think of it as a team that has but one super player playing to project him/her in the limelight at the cost of other players. Would that team win the shield? The team that collaborates with each other working shoulder to shoulder would be the winner ultimately. To foster teamwork among the staff, we need to ensure that every one on the team has a chance of getting recognition and reward as well as receiving negative reward for negative performance.

Another aspect to be noted is that the organization provides a platform for the employees to stay motivated. It includes a decent wage, working conditions, chances of career advancement, policies for fair treatment of employees and so on. Then it behooves the managers to utilize that platform to motivate the staff by providing unbiased and fair treatment, fair allocation of workload, prompt grievance handling, and so on.

Books are written on employee motivation and morale and it is not desirable to include all that material in this book. Interested readers are advised to read a good book on the subject.

Now the pitfalls into which organizations fall in the matter of employee motivation are as follows. The first pitfall is to never give any rewards or recognition. Large organizations are especially known for adopting this practice. Their argument is that the salaries include a component of the reward and yearly salary hikes are the rewards. Stability of employment is the reward. Perhaps, financial rewards may not be needed but a public recognition would go a long way in motivating a person. Administering recognition and rewards without a transparent policy on an ad-hoc basis is one other major pitfall. Another is to give the reward and recognition to the same person successively, under the justification that the individual is a super performer. Giving the reward three times consecutively would certainly demotivate all others. Another pitfall is to shy away from administering negative rewards. Negative rewards are the ones that prevent selfish working and promote teamwork.

Best practices are as follows. The organization needs to have a transparent system of recognition and rewards. The rewards, perhaps, need not necessarily be financial in nature. These are administered at regular intervals of time. These are fairly administered without any bias and based on objective data which each of the contestants can verify and agree with. It is ensured that the reward and recognition is given to as many employees as possible. No single individual receives the same reward more than twice consecutively.

14.2.6 *Quality Assurance*

Quality assurance goes a long way in ensuring that the deliverables of the requirements engineering activity are defect free in the first place and strive for excellence in the second place. Many organizations argue that quality assurance does not add quality but only verify that it exists. So it is a cost which can be avoided. The philosophy of Total Quality Management (TQM) advocates placing emphasis on the process than on inspection and testing. True enough. But look at QA as you would look at the existence of a police department. The police may not be able to prevent a crime from occurring nor can it solve every crime. But its mere existence inhibits many a prospective criminal from committing a crime. Similarly the existence of a quality assurance function improves the diligence of the performers from injecting defects into the deliverable.

Reviews are the quality control tool to verify the quality of the deliverables of requirements engineering. We have three types of reviews, namely the peer reviews, managerial reviews and expert reviews. Most organizations do implement peer reviews. Often organizations skip either the peer review or the managerial review. Expert reviews are required to validate the requirements. It is rather the exception than the practice to implement expert reviews in the organizations. Expert reviews can bridge the gaps in the requirements, which exist because the user has forgotten or the analyst has missed some aspects.

The best practice is to implement all three types of reviews for all requirements engineering artifacts.

Another important aspect is to prevent defects. This is achieved by a robust organizational software engineering process and defining/adopting international standards and guidelines such as IEEE software engineering standards. Many organizations neglect this aspect taking refuge under the argument that standards inhibit the creative instincts of the individuals. Standards ensure a minimum level of quality in the deliverables. It is always possible to take waivers when an individual comes up with an alternative that is even better than what is defined in the standards. Second, the standards are not set in stone; they are amenable for improvement. Just as organizational processes are improved, standards can also be improved by dovetailing the best practices uncovered in the projects that were already executed in the organization or out of new developments in the field.

The pitfall of organizations is to define a weak process coupled with no standards or weak standards. Another pitfall is the non-implementation of the process and standards by giving waivers to too many projects.

The best practice is to have a robust set of processes and standards and diligently improving them periodically. This set of processes and standards is implemented diligently in all projects and waivers are given under really extenuating circumstances only.

14.2.7 Knowledge Repository

A knowledge repository would consist of self-study materials to gain/update knowledge of employees, records of projects completed in the organization and any other relevant materials. Most professional organizations would have a knowledge repository. A knowledge repository would aid in effectively performing any activity by providing reference material from within and without the organization.

The pitfall many organizations fall into is not having a knowledge repository at all. Some do have a knowledge repository but only in the name without any usable material in it. Some have it, but it is poorly organized. Some organizations treat the organizational knowledge repository as a dumping ground for records of completed projects.

The other pitfall is that organizations do not update the project records to "as built" state before entrusting them to the knowledge repository. This leaves the records as they were prepared. The changes that took place during execution are not dovetailed back into the records. Referring to records that are at variance with actual occurrences is futile and provides wrong guidance to the resources referring to them.

Another pitfall is an unorganized knowledge repository. All the information is contained in the knowledge repository but it is not easy to extract the required information. One has to sift through the records manually one-by-one. This makes it tedious to use the knowledge contained in the repository.

The best practice is to consciously plan and organize an organizational knowledge repository. Update the records to as-built stage before consigning them to the repository. Normalize the metrics with actual values and subject them to variance analysis before including them in the knowledge repository. Update the information contained in the knowledge repository with state-of-the-art information. Have a set of dedicated staff that is well versed with knowledge management to diligently update the information on a regular basis. This goes a long way in ensuring that the organization moves towards excellence in the matter of requirements engineering and management.

14.3 Project Level Pitfalls and Best Practices

While the pitfalls and the best practices at the organizational level have severe impact on all projects across the organization, the pitfalls and best practices also occur at the project level. In fact, the best practices and pitfalls that occur in the project are the ones that trigger improvement of the processes and standards at the organizational level. Let us look at the pitfalls and best practices at the project level.

14.3.1 Planning

Some projects do not consider the activity of requirements engineering while planning the project. Requirements engineering is treated as a necessary evil before we can begin coding. The oft forgotten requirements engineering activity during planning is the requirements change management. When changes come in, they are treated on a case-by-case basis. This would cause what is popularly referred to as "scope creep"—increase in the amount of work that needs to be carried out. The best practice is to focus on the activities of requirements engineering while carrying out project planning and provide for resources to carry it out effectively.

14.3.2 Preparation for Elicitation and Gathering of Requirements

If we wish to capture the requirements comprehensively, preparation is essential. We need to acquaint ourselves with the domain, prepare formats, templates, ensure that all concerned personnel would be available and that only the right persons are being contacted and so on before we embark on requirements elicitation. More often than not, analysts approach this activity with little or no preparation at all. This will result in making multiple trips or not capturing the requirements comprehensively.

The best practice is to prepare well before we begin the activity of requirements elicitation and gathering.

14.3.3 Misunderstanding About Requirements

In many cases, the requirements are wrongly understood. They are understood to be only the core functionality requirements stated by the customer or end users. Ancillary functionality requirements are often missed out or left for the software designers to provide them. Most of the ancillary functionality requirements do not come from the end users or the customers. They may have to be generated by the team or from experts in the field.

The best practice is to take ancillary functionality requirements into consideration and establish them also along with the core functionality requirements.

14.3.4 Vague Requirements

It is easy to miss requirements that are not objective in nature while establishing the requirements. Requirements like "ease of use" or "aesthetically appealing" are difficult to interpret or implement. By using documentation guidelines, and effective peer reviews, we can avoid such vagueness from creeping into our established requirements. We often see such vague requirements in the established requirements. This shows lack of diligence on the part of those defining and reviewing the requirements artifacts.

Best practice is to ensure that no requirement is left vague. This can be achieved by having the right documentation guidelines, as well as, having an effective peer review.

14.3.5 Modeling Issues

Now, we have plenty of diagrams to model the data and the system. We can draw any number of diagrams to model the present system or the way as we understand the system. The diagrams are great if used properly in understanding the system. After all, a picture is worth a thousand words. But often, we draw diagrams in a complicated manner making it difficult to make any sense out of them. One, the diagrams take a lot of time to draw them and two they take a significant amount of time from those who try to interpret them and carry out the software design. Therefore, we need to use the modeling diagrams judiciously. We must draw a diagram when it aids clarity and to simplify narration. We should not avoid narration because we included a diagram. The diagrammatic tools available in software engineering are not as efficient as engineering drawings in communicating information—yet.

Therefore, the best practice is to provide a judicious mix of diagrams and narration to make our establishment of requirements in a clear and lucid manner.

14.3.6 Prioritization of Requirements

During the establishment of requirements, we ought to prioritize individual requirements. Often times, this aspect is overlooked. When we do not prioritize the individual requirements, their fulfillment would be based on the convenience of the project team rather than the necessity of the end users. This results in the situation of "urgent functionality is not yet ready and for the functionality that is ready for use end users are not ready".

The best practice is to prioritize the requirements especially from the standpoint of the necessity of the end users. This will enable us to execute the project and making deliveries that would be put to immediate use and thus save the investment made in the software project.

14.3.7 Change Management

We know that almost all software projects would have changes during the execution phase of the project. Still, it is not uncommon for projects to be negligent on the aspect of requirements change management. Even when the organizational process mandates planning and managing the change requests, the project management neglects this aspect sometimes. Planning for change management and implementing the plan should not be neglected. Careful planning for handling mid-project change requests and diligent implementation of the change management plan are best practices.

14.3.8 Tracing and Tracking of Requirements

It is not uncommon to miss out on some requirements specified by the end users. As we carry out software design, construction and testing, we miss out on some of the requirements. To ensure that we do not miss out any requirement, we use the requirements traceability matrix tool. What often happens is that we neglect to maintain the traceability matrix under pressure from various software engineering and management activities. This is the pitfall that many project managers fall into.

The best practice is to diligently maintain and update the traceability matrix at every stage. Then verifying it regularly to ensure that the requirements are all included in every stage of software engineering ensures that all requirements are included in the software product.

14.3.9 Supervision

The quality of supervision has a very significant impact on any human endeavor. In requirements engineering and management too, the same is true. Organization provides a platform for project managers to motivate their resources. Fair allocation of work, fair grievance handling, provision of adequate duration for completion of assignments, fair and equitable recognition and rewards, opportunities for learning new skills and updating of existing skills and so on would go a long way in keeping the morale high and motivates the employees. It may not be an exaggeration to state that more employee separations happen due to poor supervision than organizational policies.

The best practice is to ensure that the supervision of resources working on the requirements engineering activity is carried out fairly and equitably. It may be necessary to train the project managers in managing people to achieve fair supervision.

14.3.10 Project Postmortem

Hospitals periodically conduct a conference referred to as "death conference". In this conference, all cases of death that took place in the hospital are discussed. The pathologist who conducted the postmortem leads the discussion by giving the cause of the death and how the treatment slowed/hastened the death. The doctor who treated the patient discusses the assumptions and diagnostic decisions he/she arrived at and so on. The mistakes committed, if any, are discussed openly. It is from these conferences that all the doctors learn so that the mistakes are not repeated again. Similarly when we complete a project, we also need to conduct a project postmortem meeting and discuss the achievements and failures so that all participants learn from both the positive aspects and negative aspects of the project. Unfortunately, this is not conducted for all completed projects. When it is conducted, we discuss only the positive aspects and sweep the negative aspects under the carpet. Such meetings would not achieve any positive learning. They are a waste of time.

The best practice is to conduct a thorough project postmortem meeting and discuss all aspects including achievements and failures. It should be led by the person who conducted the phase end audit for the project closure and the project manager who managed the project. The objective of the meeting should be to learn from the completed project so that we can avoid the mistakes and take advantage of the best practices. This is a best practice.

14.4 Final Words of Pitfalls and Best Practices

We have been executing software projects in a methodical way beginning with the onset of SSADM for over three decades. We have gathered a significant amount of knowledge on this topic of requirements engineering and management. It is a matter of concern that poor requirements engineering and management continues to be the number one reason for software product failure. Unless we carry out the requirements engineering and management effectively, we may not build the "right" product or worse still, we may end up building the "wrong" product.

Therefore, it is imperative, for all of us involved in requirements engineering and management, as well as, project management, to avoid the pitfalls and adopt the best practices. This chapter is a step in that direction.

Chapter 15
REM in Agile Projects

15.1 Introduction

Agile projects use a variety of software development methodologies for developing software. All these methodologies adhere to what is known as "Agile Manifesto" which states, thus:

"We are uncovering better ways of developing software by doing it and helping others do it. Through this work we have come to value:

1. **Individuals and interactions**—over process and tools
2. **Working software**—over comprehensive documentation
3. **Customer collaboration**—over contract negotiation
4. **Responding to change**—over following a plan

That is, while there is value in the items on the right, we value the items on the left."

There are twelve principles behind Agile Manifesto:

1. Our highest priority is to satisfy the customer through early and continuous delivery of valuable software.
2. Welcome changing requirements, even late in development. Agile processes harness change for the customer's competitive advantage.
3. Deliver working software frequently, from a couple of weeks to a couple of months, with a preference to the shorter timescale.
4. Business people and developers must work together daily throughout the project.
5. Build projects around motivated individuals. Give them an environment and support they need, and trust them to get the job done.
6. The most efficient and effective method of conveying information to and within a development team is a face-to-face conversation.
7. Working software is the primary measure of progress.
8. Agile processes promote sustainable development. The sponsors, developers and the users should be able to maintain a constant pace indefinitely.

M. Chemuturi, *Requirements Engineering and Management for Software Development Projects*, DOI: 10.1007/978-1-4614-5377-2_15,
© Springer Science+Business Media New York 2013

9. Continuous attention to technical excellence and good design enhances agility.
10. Simplicity—the art of maximizing the amount of work not done—is essential.
11. The best architectures, requirements and designs emerge from self-organizing teams.
12. At regular intervals, the team reflects on how to become more effective, then tunes and adjusts its behavior accordingly.

This is available on www.agilemanifesto.org.

Another hallmark of agile methodologies is that they consider requirements as always in an emerging state and that they are never finalized. So changes can occur at any time. The software design is aimed at fulfilling the present requirements and not for some assumed future requirements. Agile methodologies normally have smaller teams and smaller products even though we now find that agile is being used for larger systems as well.

I am neither a fanatic-fan nor an antagonistic-critic of this methodology. I do not attempt to either criticize or extoll the agile methodology. I am just presenting it as information to you, the readers.

What I believe is that agile methodology, just as other software development methodologies, has its place and works best in certain scenarios and may not be effective in other scenarios. I also believe that this is neither the first methodology nor the ultimate panacea for software development.

There are many methodologies considered to be conforming to the agile manifesto, of which the following are the popular ones:

1. XP (Extreme Programming)
2. Scrum
3. DSDM (Dynamic Systems Development Method)
4. Feature Driven Development
5. Test Driven Development
6. Adaptive Software Development
7. RUP (Rational Unified Process) and AUP (Agile unified Process)
8. Kanban
9. Crystal Clear.

Let us discuss how requirements are handled in each of these methodologies. I am including a very brief description of each of these methodologies. To include a comprehensive treatise about each of these methodologies would take too much space. Interested readers may read specialized books on the methodology that interests them. The focus here is on the requirements management in each of these methodologies.

Another important aspect of agile methodologies is that the organizations implementing these methodologies can tailor the actual implementation to suit their unique culture. Therefore, there would be many variants in vogue of what is described in this chapter.

15.2 Extreme Programming

Extreme Programming (XP) goes through six phases, namely the **exploration, planning, iterations to release, putting into production, maintenance and death**.

In the exploration phase two main activities are carried out. The customer writes the story cards as the first activity. Each card describes a feature and all cards put together include all the requirements for the first release of the software. The second activity is carried out by the developers, who familiarize themselves with the development environment.

In the planning phase, the developers prioritize and schedule the development for the first phase. The schedule is usually limited to 2 months.

In the iterations to release phase, multiple iterations may take place for each of the releases. Normally no iteration would exceed 4 weeks duration. The development team selects the stories for development in consultation with the customer.

The putting into production phase would see that one release of the software is released to the customer after a final round of testing. If the customer asks for any changes in the software, they will be implemented. The release process could be in iterations, each of which would normally be limited to a 1 week duration.

After putting one release into production, the team would take up the next release of the proposed software product.

In the maintenance phase, the development team would support the customer in effectively utilizing the released software. Sometimes new people could be inducted into the team for carrying out the maintenance.

The death phase begins when the customer has no more stories that need development of software. It also requires the stable operation of the software that was put into production, and the development team is no longer required to support the system in production.

As you can see, the requirements are handled in the exploration phase and the putting into production phase. In the exploration phase, the customer writes the user story cards. Each card would contain one feature of the software. Any shortfalls, ambiguities or vagueness in the story cards is resolved using a face-to-face communication with the customer who is co-located with the development team. During the putting into production phase, the customer may request changes in the developed software, using the face-to-face communication.

The customer is chiefly responsible for defining the requirements for the proposed project. The programmers develop programs realizing the user stories. Programmers and testers can take clarifications from the co-located customer about the user stories whenever they require clarification.

15.3 Scrum

Scrum is a term taken from the game of rugby. When the ball is not in anybody's hands and both teams struggle to get it crowding around the ball, the term used to describe the scenario is "scrum-mage". It is generally taken to mean "a brief and disorderly struggle or fight". Scrum is steadily gaining more popularity among agile methodologies and a certification (Scrum Master) is also offered for proficiency in using Scrum methodology.

Scrum manages the project in three phases, namely the **pregame phase**, **development phase** and **postgame phase**.

Pregame phase includes two main activities, namely the planning and the architecture design. During planning, a "product backlog list" is created. This list would consist of all the product requirements, to the extent possible. This list would be owned by the "product owner" who is a team member. Now, how the list is actually s created may vary from organization to organization. In some cases, the product owner would elicit, gather, analyze and establish the product backlog. In other cases, the list would be filled in by the customer, the marketing department, the field support personnel and the developers themselves. Another notable feature of Scrum is that this product backlog list can be constantly updated including addition of newer items. But the development team interacts with only the product owner for all matters concerning the project requirements. The planning phase includes prioritizing the requirements. Planning also includes defining the development environment, risk assessment, progress control, training and so on. As part of the architecture definition activity, the high level design of the proposed product based on the enumerated product backlog is created.

The development phase consists of "sprints". A sprint is an iteration of software development that results in the release of a portion of the proposed software product fulfilling a part of the overall functionality. Each sprint can include requirements analysis, software design, construction and delivery. The duration of a sprint is usually restricted to less than 4 weeks. This phase, in sprints, sees the development of the software product fulfilling all the requirements. Requirements change management also takes place in this phase. By the time this phase is completed, all requirements would have been met by the software product.

When all the requirements are implemented in the developed software product, the postgame phase begins. Overall product testing may also take place in the postgame phase, if necessary. The postgame phase includes handing over the system and supporting the customer in the usage of the delivered system.

The requirements are primarily defined in the planning phase and to a lesser extent in each of the sprints. The product owner champions the establishment of the product backlog list. The customer and other stakeholders do assist the product owner in the establishment of the product backlog list. Any stakeholder intent on modifying any of the requirements would interact with the product owner and modify the product backlog list.

The product backlog list is the set of requirements defined for the project. This list could include product features, and functionality in fresh development projects. It would include defects, and enhancements for a software upgrade project.

The backlogs for a sprint are identified during a "sprint planning meeting". The product owner, users, customers, organizational management, and Scrum team participate in the planning meeting. The meeting decides the functionality selected for realization during the next sprint. The users, customer and management participate in the first planning meeting and all the subsequent meetings are only attended by the product owner, and the Scrum team.

15.4 Dynamic Systems Development Method

Dynamic Systems Development Method (DSDM) makes the philosophy of software development stand on its head! Instead of freezing the functionality for the proposed product and scheduling it, it freezes the duration and then selects the functionality to fit the duration!

DSDM is carried out in five phases, namely, **feasibility study**, **business study**, **functional model iteration**, **design and build iteration**, and **implementation**. The phases of the feasibility study and business study are carried out only once and the remaining three phases are iterated. Each iteration is referred to as a "time box" in the DSDM taxonomy. Each time box is planned with a set of functionality and a fixed duration with the objective of fulfilling the set functionality within the time box. Usually, the maximum duration for a time box is a few weeks not exceeding two calendar months.

During the feasibility study, the feasibility of executing the project using the DSDM is assessed based on the type of project and the organizational culture. The phase deliverables are a feasibility report and an outline plan for development. The feasibility report would include details of technical feasibility and the associated risks. If the risks are major, then a prototype of the proposed product may also be built. The outline plan will include schedule of the project with major milestones and resource requirements.

During the business study phase, the knowledge transfer takes place from the stakeholders to the development team. This is advocated to be achieved by conducting meetings or workshops. These would consider all aspects of the proposed system and set development priorities. The business processes are understood by the development team and sometimes are even documented. The deliverables from this phase are business area definition, system architecture definition and outline prototyping plan. The business area definition describes the business processes in a high level manner. The system architecture definition is a sketch and it is expected to change during the course of the development. The prototyping plan would consist of the strategy for prototyping and the configuration management plan.

In the functional model iteration, a prototype for the iteration is designed, using which the product would be developed. A functional model (use and improve prototype) is built in this phase which would be handed over for the next phase along with the source code. The other deliverables from this phase are the prioritized functions, functional prototyping review documents, non-functional requirements (or ancillary functionality requirements), and risk analysis for the next steps of development.

The design and build iteration phase sees the development of the proposed software product. The deliverable of this phase is a tested system that fulfills a set of functionality earmarked for the iteration.

The implementation phase is where the system is transferred to the production system. The product is implemented, training to users is conducted, and the software is rolled out into production including handholding of users during initial usage of the implemented system. In addition to the implemented system, a user manual and a project report are the deliverables of this phase.

The last three phases are iterated until all functionality is built into the proposed software product and it is implemented on the production system.

The project requirements are collected during the business study phase and are documented in the business area definition document. Further, DSDM proposes three roles for handling the project requirements, namely, the **Ambassador User**, the **Adviser User** and a **Visionary**. The Ambassador User is normally one of the users of the proposed system. The Ambassador User is the main interface between the development team and the users, and brings the domain knowledge to the team. An Ambassador User provides clarifications to the development team as and when requested. An Ambassador User also provides the project progress to all the project stakeholders. An Advisor User is similar to an Ambassador User, but has some special expertise in the proposed system. He may be a member of the IS department bringing in the operations point of view to the team or an auditor who brings security aspects to the team and so on. A visionary is the one who provides the business objectives of the proposed product and the project. Visionary is usually the one who came up with the idea for the development of the proposed software product.

15.5 Feature Driven Development

Feature Driven Development (FDD) develops software in five phases, namely, **develop an overall model, build a features list, plan by feature, design by feature** and **build by feature**. FDD assumes that requirements are already established before the project begins and takes off from that point onwards.

During the "develop an overall model" phase, the domain experts make a presentation to the development team and walk them through the functionality. In the case of large systems, the functionality may be further divided and multiple presentations are made to the development team. The objective of this walk-thru meeting is to familiarize the development team with the functionality for the proposed software product. The development team discusses the functionality and constructs an appropriate object model for the proposed software product.

In the "build a features list" phase, the development team draws up a features list which would contain "client-valued" major features for the proposed software product. Each of the major features may further be subdivided into features. The feature list prepared by the development team may be reviewed by the users and the domain experts to confirm that all required functionality is included.

In the "plan by feature phase", an overall plan is prepared in which the features are prioritized. Then based on the dependencies between the features, sets of features are assigned to "chief programmers". A schedule for developing the assigned features is also drawn up.

The phases of "design by feature" and "build by feature" are iterated until software is developed for all features. In these two phases, a set of features is selected, then designed and developed. Each iteration is normally limited to 2 weeks or less. These two phases include, design, design verification, coding, unit testing, integration and code walk through. Upon completion of an iteration successfully, the developed code is included in the build of the main product.

This methodology assumes that requirements already exist when the project begins. These are converted into features list through the collaboration between the domain expert and the development team. These features drive the design and development of the proposed software product. As FDD is also an agile method, it ought to accept changes during the two phases of designing and building the product.

15.6 Test Driven Development

Test Driven Development (TDD) is based on the premise that testing is crucial to developing a defect-free product. To be able to write any test case, one should know the functionality as well as the software design. To write a test case for unit testing, in addition to knowing about the functionality and the design, even the knowledge of code is also necessary. Thus, TDD forces one to think through the project requirements and software design.

The phases in TDD are, **write/add an automated test case**, **run the test** (and it fails), **write the production code** and run the test again, and when the code passes the test, **refactor the code**. Then the next cycle begins by adding another test case.

By test case, I mean it is a test case for conducting a unit test. Adding an automated test case requires using an automated software testing tool, and we need to write some code to be able to run the test case. That involves learning the functionality, and designing the component before we can write some rudimentary code. But we write an efficient unit test case. As we write test cases, we would have a set of robust unit tests and it will result in a quality product because the unit test is the most crucial test in achieving software quality.

During the phase of writing production code, we improve the preliminary code to make it robust. Then we run the automated test case once again. Writing/ modifying the code and re-testing it can go through multiple iterations until the code passes the automated test case comprehensively.

During the refactor phase, the code that passed the test would be cleaned by removing all trash code and unused variables.

Now the cycle repeats until all functionality is achieved and it is no more possible to write any new automated test cases.

TDD methodology does not include establishment of requirements. TDD assumes that requirements are already established before the development starts. Handling change requests involves changing the automated test cases. It has to be done as and when change requests are received.

15.7 Adaptive Software Development

ASD or Adaptive Software Development is also an incremental and iterative development methodology. That is, the software is developed in increments and each increment is developed in iterations. That way, it is possible to handle large software projects under this methodology. ASD also encourages the use of the use-and-improve prototyping as the main software engineering technique.

ASD goes through three phases, namely, **speculate**, **collaborate** and **learn**. The term "Speculate" is used to highlight the aspect of planning; "Collaborate" is used instead of development/implementation/realization/construction to emphasize team work; "Learn" is used to acknowledge mistakes already committed and to learn from them.

The Speculation phase consists of two activities, the project initiation and the cycle planning. During project initiation, three artifacts are prepared, namely, the project vision charter, project data sheet and product specification outline. The project vision charter contains the project objectives in brief. The project data sheet is usually one page containing vital information about the project. The product specification outline again is a brief statement of the functionality for the proposed software product. In the cycle planning, the team estimates the duration and resources required to complete the cycle. In each cycle, a set of components are engineered and constructed by the team. During the learning phase, the quality control activities are carried out and the software components developed in the cycle would be released. The learning from the mistakes uncovered during the quality control activities are plowed back into the next cycle planning.

ASD as its name suggests does not prescribe any roles and responsibilities or software development methodologies, but adapts these from the existing practices of the organization. Each organization can implement its own practices and management style in the project execution.

ASD assumes that requirements are established. No one on the team assumes responsibility for the requirements establishments. The customer is expected to be co-located with the team and therefore, we need to assume that the customer would explain the requirements and guides the team in achieving the desired functionality for the proposed software product.

Changes in requirements are assumed and are implemented as and when a change is received in the collaboration phase.

15.8 RUP and AUP

AUP (Agile Unified Process) is a scaled down version of RUP (Rational Unified Process) of Rational Corporation, now a part of the IBM.

RUP itself is an iterative process with four phases, namely, **inception, elaboration, construction** and **transition**. All of these phases are iterated until all software is completely developed.

In the inception phase, the scope of the project is defined, the functionality earmarked for the iteration is defined through use cases, acceptance criteria is defined, initial software architecture is devised, and the schedule and cost are estimated.

In the elaboration phase, software design is carried out, tools for use in the project are identified, an executable prototype is created, and the development environment is defined and created. Once this phase is completed the team is ready to embark on developing the proposed software product for the iteration.

During the construction phase, all software development is carried out for the iteration, including normal testing. Any changes requested are also implemented. This phase ensures that all project objectives of productivity, quality, costs and schedule are met. Completion of this phase sees the functionality set for the iteration is realized in the software product and it will be ready for release.

In the transition phase, the constructed software is released to production use of the end users in the actual environment. In this phase, the activities of beta testing, piloting, end user training, documentation like the user manual, operations manual, troubleshooting manual, etc., are prepared, and rolling out the product into production is carried out. Handholding of end users during initial usage is also part of this phase. Normally two types of software release are carried out in this phase, namely, the beta release and general release. Beta release is for testing by the end users and general release is for putting the software into production.

RUP uses nine workflows namely, **business modeling** (modeling of the business process proposed for the project), **requirements, analysis and design, implementation** (software development), **test, deployment, configuration and change management, project management** and **environment** (software development environment including tools, and other support to development).

Requirements are gathered in the inception phase and are developed further in the elaboration phase. The requirement activity is carried out by the Business Process Analyst. The requirements are captured using the use cases. A Business

Model Reviewer reviews the defined use cases to ensure quality in the defined requirements.

While RUP is an iterative development model, it is not considered an agile method. Its variant AUP (Agile Unified Process) is used as an agile method for software development. The commonality between both models is the dependence on use cases as the method for capturing requirements and software design.

AUP uses the same four phases as in RUP, namely, the **inception**, **elaboration**, **construction** and **transition** phases detailed above. It also emphasizes the iteration of these four phases to realize all the user requirements in the proposed software product.

AUP utilizes only seven workflows instead of nine workflows of RUP. These are, **modeling**, **implementation**, **test**, **deployment**, **configuration management**, **project management**, and **environment**. The requirements workflow is merged into the modeling phase. Architecture definition part of the analysis and design phase is merged into the modeling phase. The detailed design part of the analysis and design phase is merged into the implementation phase.

AUP uses two types of software release, namely the development release and the production release. The development release is for deployment on the target system and is used for quality control and training the end users. The production release is the software ready for putting the system into production.

The requirements in the AUP are established utilizing the use case method in the modeling phase. Change management happens throughout the project. The Business Analyst takes responsibility for defining the requirements and tracing them through the software development.

15.9 Kanban

The Kanban technique has originated in the Japanese manufacturing plants of the Toyota Corporation. Kanban in Japanese means "signboard/signaling device". This technique was utilized to achieve JIT (Just In Time) manufacturing. In high-volume manufacturing the practice before the onset of Kanban, was used to produce the maximum number of components at each of the workstations. This has led to a pile up of components at a few workstations that had higher capacity than the downstream workstations. Over a period of time, the pile up was significant, locking up scarce capital. To reduce the pile up of components at workstations, the Kanban technique was introduced. Kanban suggests that a new batch of components should be started if and only if the previously produced components are picked up by the downstream workstation. Kanban has changed the concept of "pushing" to "pulling".

"Pushing" indicates pushing components from an upstream workstation to the downstream workstation without considering whether the downstream workstation is ready to receive them.

"Pulling" indicates a "downstream workstation pulling the component from the upstream workstation" and the upstream workstation would produce components only when the previous batch of components is picked up by the downstream workstation. That is, the upstream workstation produces components "just in time" for the downstream workstation to begin its operation. This has resulted in huge savings for Toyota and it has been adopted at most high volume manufacturing organizations.

Now software development organizations also are trying to adopt this technique into software development field. Releasing the entire software in one big-bang would result in it being idle if the implementation parameters are not yet made ready. So, it advocates delivering software in iterations to those sites which are ready for implementation and use in production.

The Kanban philosophy of software development focuses on delivering the software just-in-time for implementation. In many organizations, software is developed well in advance of installing the IT infrastructure making the software wait for the facility to come up, hardware to be received, system software to be received and so on. This is locking up valuable capital. So Kanban places emphasis on scheduling the software development work "to be ready in time" rather than on "as soon as possible" basis.

The Kanban system does not prescribe any software life cycle or development method. Kanban advocates an iterative development methodology to deliver software in installments. The development organization can follow their existing software development processes.

There are five steps in Kanban development, namely, **visualize the workflow**, **limit WIP** (Work In Progress), **manage flow**, **make process policies explicit**, and **improve collaboratively**.

To visualize the workflow, a set of cards are pinned to a wall referred to as the card-wall. The cards on the wall are arranged in columns. Each card column would contain the steps of a process flow of the proposed system.

Limit WIP advocates against taking up too many modules in parallel. It suggests taking up one or two modules at a time so that workable software becomes ready faster. If ten team members take up ten programs of ten different modules it will make the WIP higher. On the other hand, if ten programmers of the team take up ten different programs of one or two modules, there will be something ready for use in a short time.

Manage the flow indicates that the workflow is measured, reported and monitored at each step. This will ensure that the iteration is executed efficiently and the iteration delivery is ready on time.

Making the policies explicit, involves every stakeholder in the manner the work is carried out by the team. That way, the stakeholders can give suggestions for improvement and participate in discussions effectively. This would make it easier to arrive at a consensus easily.

Improving collaboratively involves allowing suggestions for improvement to come from any source. All suggestions received are analyzed by all stakeholders and agreed suggestions are implemented incrementally in the process steps.

The requirements are captured in the cards and pinned on the card wall. The column would be maintained on the wall until its functionality is realized and will be removed on the release of the software. Changing a requirement means changing a card on the wall. The tracing of requirements through software development would depend on the specific software development methodology implemented in the project.

15.10 Crystal Clear

Crystal methodologies believe that more rigor in software development is required as the project size increases. They consist of four methodologies coded by color, namely, Crystal Clear (no color), Crystal Yellow, Crystal Orange and Crystal Red. A Crystal Clear project would have about 6 team members, a Crystal Yellow project would have around 20 team members, Crystal Orange project would have about 40 team members and a Crystal Red project would have about 80 team members. As can be seen, a Crystal Red project would need the maximum rigor. The rigor needed would progressively diminish in Orange, Yellow and Clear projects.

There are some common aspects in all Crystal project types. They all utilize an incremental software development life cycle. Each increment is limited to a duration of four calendar months with most of the iterations falling between one and three calendar months. Face-to-face communication is the preferred method of communication for establishing requirements or requesting changes. The customer or customer representative is expected to be co-located with the development team.

Of these, the Crystal Clear methodology is considered to be agile. Crystal Clear methodology consists of 6–10 team members. Crystal Clear has the following properties:

1. Workshops for transferring domain knowledge to developers
2. Incremental development life cycle with frequent deliveries of usable software
3. Osmotic communication, by customer being co-located with the development team, or in other words, close interaction with users
4. Advocates automated tests, at least, in regression testing
5. The iteration is to be limited to a maximum of three calendar months.

Crystal clear insists on testing tools, configuration management tools and use of white boards in place of documents. White boards would contain the design of the proposed software as well as meeting summaries.

The crystal method uses eight practices in each increment to develop software. These are **staging**, **revision and review**, **monitoring**, **parallelism and flux**, **holistic diversity strategy**, **methodology-tuning technique**, **user viewings**, and **reflection workshops**.

During staging, the next increment would be planned. The team selects the requirements they can deliver in three calendar months and schedules the work.

Revision and review accomplish the code development/modification and review of the objectives of the increment. These two activities are carried out in all the iterations.

Monitoring involves monitoring the increment progress. The monitoring is carried out by means of reaching the milestones, defined in the staging step. The result of monitoring may be that the product is mature for reviewing or it is still in development.

Parallelism and flux indicate that teams can work in parallel, especially in Crystal Orange or Red projects. When working in parallel, the synchronization becomes important. Careful monitoring would ensure synchronization of work between parallel teams.

Holistic diversity strategy splits large functional teams into cross-functional groups so that each team would have specialists from multiple disciplines. This allows the project to have smaller teams with required specialties. Instead of having specialists at a central place and being used by all teams on a need-basis, they are allocated to the team itself.

The methodology tuning technique uses team workshops and meetings to select a development methodology and tune it to the needs of the project. In each increment the team can draw lessons from the present increment and use it to improve the performance for the next increment.

The user viewing techniques are basically reviews by end users ahead of the software release. Crystal recommends two user viewings per increment. This ensures that all user requirements are met by the release.

The reflection workshops are held twice: the pre-increment reflection workshop and the post-increment workshop. Pre-increment reflection workshops facilitate reflection on the experience of the previous increments so that the present increment could be executed efficiently. The post-increment workshops facilitate reflection of the events of the increment and to draw lessons for the future increments.

Crystal methods do not prescribe any development life cycle. So, the team can choose any life cycle. Since there are no set phases, requirements are not restricted to any one phase. However, the staging activity would have to consider requirements. Normally Crystal clear uses informal use cases to record requirements. These would be traced through the software development using release plan and test cases. Normally a user manual is also prepared and it would also contain proof that requirements are indeed implemented.

15.11 Establishment of Requirements in Agile Projects

All agile projects do have some common aspects:

1. They all believe in the agile manifesto
2. They all require the full involvement of the customer or customer-representative in the software development. Most specify co-locating the customer with the

project team. Some of them go to the extent of mandating that the customer/
representative/user must be part of the project team
3. They all believe in developing the code
4. They try and minimize the documentation
5. They all focus on delivering the code, that is the software engineering part
6. They all do not talk about the "management part" of the software development
   project.

All in all, strictly speaking agile methods are software engineering methodologies rather than management methodologies. Second, they all focus on designing,
building and testing the product and more or less silent on the aspects of
requirements engineering and management. They also do not acknowledge the
need to trace the requirement through all the engineering artifacts.

The philosophy of agile methodologies seems to be that the customer is
responsible to take care of the requirements and ensure that they are implemented.
By co-locating the customer or making the customer part of the project team they
ensure that the requirements are efficiently taken care of. As, many projects are
handled using the agile methods, we have to acknowledge that the philosophy is
indeed working.

Still, agile projects do establish the requirements in some way even though the
rigor of such establishment may not match with the rigor of the other software
development methodologies. Here are some ways in which requirements are
established in agile projects:

1. **User Story cards**—users write user stories on cards. Each card would contain
   one feature for a program. One card may result in one program or multiple
   cards may be necessary to complete a program. The card may not give full
   information to the programmer. More often than not, a conversation between
   the programmer and the user would be necessary for the programmer to fully
   comprehend the requirement and program it.
2. **Task list**—Task list is another means to document the requirements. It is an
   enumeration of the requirements or user stories that need to be developed. This
   list gets updated as and when the tasks are completed to indicate the project
   progress.
3. **CRC cards** (Class, Responsibility and Collaboration)—Each card records the
   requirements and the interaction between requirements in such a way that
   software can be designed and constructed.
4. **Customer acceptance test cases**—List of user acceptance test cases also serve
   as project requirements.
5. **Feature lists**—List of features to be built into the product is another mechanism to establish project requirements. Each feature must possess some client
   value and should be implementable in 2 weeks or less.
6. **Informal Use Cases**—Agile projects use scaled down use cases with much
   reduced rigor in describing the scenarios of the use cases
7. **Product backlog**—This is another mechanism to record the requirements. In
   this the features yet to be completed are enumerated. These are described not as

requirements or features but as pending items or backlogs of functionalities that are yet to be completed.

8. **Test Cases**—Test cases are developed in place of requirements. Code is developed to pass these test cases.

9. **URS**—yes, some agile projects do use good old URS to establish the project requirements. These come in handy when the projects are outsourced to a far off location offshore, the client organization supplies the URS as customer co-location with the team is out of the question. Daily conference calls either with the video or audio are used in place of the daily standup meetings.

15.12   Tracing and Progress Monitoring of Requirements

Progress monitoring is mostly face-to-face. Normally daily standup meetings are the means by which all the project progress is communicated to all the team members of whom the customer or customer representative is also one party. It is usually the responsibility of the customer to communicate the progress of the project to any other stakeholder. And if any other stakeholder wishes to communicate with the team, the stakeholder may participate in the standup meeting or communicate the same to the customer representative who would in turn interact with the project team and resolve the stakeholder concern.

Tracing of the requirements in the product is again incumbent on the customer representative on the project team. Since a customer representative participates in all meetings of the project team, he/she can ensure that all project requirements are indeed met by the project deliverables. A formal traceability matrix is neither defined nor updated.

The following are some of the means utilized by the agile projects to trace the requirements and monitor the progress of the project.

Standup meetings—In this meeting, all the team members meet usually at the beginning of the shift for a very short duration of about 15 min every day. During the meeting, every team member informs the others about what was completed the previous day, the plan for the current day and the problems or concerns the individual is facing. These are discussed and consensus is arrived at immediately. If any issue could not be solved immediately, the manager/coach/mentor would find a solution and conveys it to the concerned team member and also perhaps in the next standup meeting. Communication of the progress to the other stakeholders is the responsibility of the customer representative.

Visible wall graphs—The progress of the project is depicted in appropriate graphs as dash boards. The color green is used to indicate an activity that is under control; red is used to indicate activities that are out of control; and yellow is used to indicate activity that is tending towards slippage.

Fig. 15.1 Example of a
burn-down chart

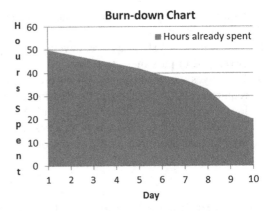

Burn-down charts—it is a line graph connecting the number of hours spent on the project every day. Burn-down chart show progress by depicting the number of hours already spent. Figure 15.1 depicts a typical burn-down chart.

15.13 Final Words on REM in Agile Projects

Agile projects shun formality and excessive documentation. Therefore, it does not make use of pre-defined formats or templates and therefore, is not presented in this book. Each organization or for that matter, each project can devise its own formats and templates to suit their unique project needs.

The practice of requirements establishment, tracing and management vary from organization to organization and perhaps project to project. Freedom to the development team is the mantra of the agile methodologies.

Appendix A
Documentation Guidelines

A.1 Introduction

The language of English or perhaps every language, for that matter, is especially rich in vocabulary with many words with almost the same meaning except for subtle variations. The grammar allows free-flowing writing and in a poetic manner. It is possible to write grammatically correct language yet obscure the real meaning behind the write up. Well, poetic language is best for writing poems and flowery language is fit for writing novels. But if we use those features of the language for capturing and recording requirements for a software project, it would be very difficult to proceed with the next steps of the project. Again, such language would not allow us to ensure that the requirements are met because of the ambiguity.

In requirements specifications, we need to use the language to mean precisely one thing that is interpreted by all stakeholders in the same way. Therefore, we need to restrict the freedom of individuals involved in requirements engineering work in documenting the requirements and provide them guidelines so that all those involved in requirements engineering would document the requirements specifications in a similar manner. Every organization ought to have a set of documentation guidelines for business writing and most professional organizations would have such documentation guidelines.

Here is a suggested set of documentation guidelines and these can be adopted in your organization as it is or with modifications to suit your unique needs.

A.2 Documentation Guidelines

This guideline covers primarily the method of presentation, composition and editorial practice to be followed in the preparation of requirements engineering documents.

M. Chemuturi, *Requirements Engineering and Management for Software Development Projects*, DOI: 10.1007/978-1-4614-5377-2,
© Springer Science+Business Media New York 2013

A.2.1   *Formatting*

All documents shall be formatted as follows:

1 The page size of all documents shall be 'letter'. If a special paper size is needed, a waiver is required.
2 Each document shall have a header except on the title page. It shall have:

2.1 The name of the document
2.2 The page number. The page numbering shall start from the table of the contents page which will be '2'

3 Each document shall have footer. It shall have:

3.1 Copyright information
3.2 The version number

4 The margins shall be:

4.1 Left margin shall be 1"
.2 Right margins shall be 0.75"
4.3 Top margin shall be 0.5"
4.4 Bottom margin shall be 0.5"

5 The typing shall be flush with the left margin. No tabbing is required for the first line in a paragraph.
6 The font size shall be:

6.1 The font shall be Times New Roman. This can be deviated from in customer documents and the documents that are sent to customers. In those documents, the font preferred by the customer shall be used.
6.2 Normal textual matter shall be of 11 points
6.3 All captions shall be bold-faced
6.4 All headers will be of 14 points size
6.5 For diagrams prepared inside the organization, the font size shall be 11 points
6.6 If the diagrams are copied from elsewhere, the restrictions of the type of font and the size would not apply.

A.2.1   *Title Page*

All documents except those that are less than three pages in length will have a title page. The contents of the title page are:

1. The name of the document in font size 16 shall be placed about one third distance from the top
2. The name of the document shall include the name of the document in the first line and the second line can have the name of the customer / the project for which this document is prepared

Table A.1 Revision History Block

Version Number	Details of Changes	Prepared by	Approved by	Date of Approval

3. The name of our organization and the month and year of preparation of the document shall be placed just above the revision history block. These will be two lines in font size 14
4. There will be a revision history block at the bottom of the page. The revision history block shall be as depicted in Table A.1.

A.2.3 Table of Contents

All documents with 5 sections or more shall have a table of contents page. It will be on a separate page. It shall be prepared using the table of contents feature and shall be set to "formal" type. It shall be the next page after the title page.

A.2.4 Content Pages

Content pages shall start on a fresh page after the table of contents page and will continue till the end of the document.

A.2.5 Terminology

All the technical terminology will conform to IEEE standard 610 "IEEE Standard Glossary of Software Engineering Terminology". If a customer enforces a different standard, a waiver needs to be obtained for using a different set other that IEEE standard 610.

A.2.6 Abbreviations

All abbreviations need to be aligned with internationally understood abbreviations. When there is room for doubt or if the abbreviation is not found in any international standards, the full form is to be given within brackets the first time the abbreviation is used in the document. The following guidelines shall be followed when using abbreviations:

1. Abbreviations, in general will be used without a full stop after them except in cases where the abbreviations result in common English words, such as:

   1.1. 'No.' for 'numbers'
   1.2. 'Fig.' For 'figure'
   1.3. 'Bull.' For 'bulletin'

2. Abbreviations shall not be used where the meaning is likely to be obscured. In cases of doubt, words should be spelled out in full.
3. Abbreviations—'e.g.', 'i.e.' and 'viz' shall not be used. Instead, the words, 'for example', 'that is' and 'namely' shall be used in their places respectively.
4. The same abbreviation shall be used both for singular and plural words
5. Letters of abbreviation shall neither be spaced nor punctuated.

   5.1. Incorrect—I S O /I.S.O.
   5.2. Correct—ISO

6. Generally, abbreviations shall not be used in main titles.

A.2.7 Definition of Terms and Abbreviations

When definitions of terms and abbreviations are included in the document, they will be preceded by words **"For the purpose of this document, the following definitions will apply"**. Such definitions are listed in the alphabetical order. Definitions of terms shall be unambiguous, precise and given in descriptive form.
   Example:

Estimate—An Estimate is a best guess of the resources required to perform a future activity.
BFS—Business Function Specification.

A.2.8 Paragraphing and Section Numbering

The text of a document shall be suitably numbered and subdivided in accordance with the method described below.
   Decimal notation shall be used for numbering paragraphs and sections. Hindu-Arabic numerals shall be used for such numbering. The scheme of numbering is shown below:

A.B.C.D.

   Where

- A—Section Number: shall be incremented for every succeeding section
- B—Paragraph number: shall be incremented within the section for every succeeding paragraph

- C—Sub-Paragraph Number: shall be incremented within the paragraph for every succeeding sub-paragraph
- D—Sub-sub-Paragraph Number: shall be incremented within the sub-paragraph for every succeeding sub-sub-paragraph

The number of levels shall not go beyond four levels.

A.2.9 Appendices

Any lengthy matter, which would not add value in the body of the document, but is of referential value in the use of the document, shall be given as an appendix.

As far as possible, appendices shall be avoided.

A.2.10 Enumeration

It is preferable to enumerate items in a bulleted form than to run them along. However, if the number of items is three or less, they may run along in the sentence itself unless for some reason bulleting is preferred. When the enumeration runs along in a sentence, each of the items is suffixed with a comma and the conjunction 'and' shall precede the last item.

When bulleting is used to enumerate items, the enumeration shall be preceded by such introductory words as, 'consisting of', 'as follows', 'conditions are', etc. and also by a dash (–). No bullet list shall stand-alone without any introduction.

Bullets shall be numbered, using numerals. For subsequent levels bullets, decimal notation as explained for paragraph numbering will be used. Example—If the first level bullet starts as 1, the second level bullet will be 1.1; the third level bullet will be 1.1.1 and so on.

Each bullet shall be confined to enumerate one single item or idea. Running enumeration shall be avoided as far as possible in bullets. When an idea is enumerated on a bullet, it shall be restricted to one or two sentences.

A.2.11 References

Care should be taken to avoid references to material that can change in time. However, exceptions to this guideline are international and local standards and organizational process documents and when these documents are referenced; their version number will also be included in the reference. When such reference is made, such introductory words as, 'conforming to', 'adhering to', 'in accordance with', 'as prescribed in' and 'as given in' shall precede its designation. If this reference is only for information, the word 'see' shall be used before the designation. The words 'as per' shall be avoided when quoting from a document.

All matter from the standard document that has crucial importance will be included in the body text itself.

A.2.12 Tables

Tables shall be used to show relationships clearly and wherever tabular presentation would eliminate repetition.

Each table shall have a caption placed at the top of the table. The caption shall contain a table number and title. The table number shall consist of the section/chapter number followed by decimal point and a serial table number within the section/chapter. Suffixing of table number with alphabets such as A, B shall be avoided. Correct example—Table 5.2—URS Template.

Each table shall as a general rule, be placed after the first reference to it and as near to it as possible, without needlessly breaking into the middle of a paragraph. However, tables requiring frequent reference from multiple locations of the document may be given at the end of the document.

In no case shall an entry in any cell be blank. Where the absence of information is to be indicated, it may be so indicated by writing 'NIL' in the cell. Where the requirement is not relevant and does not apply, it may be so indicated by writing 'NA' (Not Applicable) in the cell.

In general footnotes to tables shall be avoided. If it is imperative to use foot notes, they shall be placed immediately below the table. To indicate references to foot notes, asterisks may be used. In case there are a number of footnotes, super-scripted numerals in one consecutive series shall be used.

Other aspects to be considered are:

1. Align digits from the right. Use decimals rather than fractions and align decimal points
2. Do not use ditto marks
3. Symbols such as # and % may be used only in column headings to conserve space
4. Place currency sign, where necessary, in the column heading and with the totals.

A.2.13 Illustrations

Illustrations, such as, line graphs, bar charts, and pie charts shall be used wherever it is possible to illustrate an idea more clearly, concisely and accurately.

A.2.13.1 Captions for Illustrations

Each illustration shall have a caption, placed at the top of the illustration. The caption shall contain illustration number and title. The caption shall be prefixed with word—'Fig.' The illustration number shall consist of the section/chapter number followed by decimal point and a serial illustration number within the section/chapter. Suffixing of illustration number with alphabets such as A, B shall be avoided for stand-alone illustrations. However, where illustrations are grouped together and presented at one place, the illustration number may be prefixed using alphabets A, B, C, etc. The first letter of all the principal words in the caption shall be in capitals.

Each illustration shall as a general rule, be placed after the first reference to it and as near to it as possible, without needlessly breaking into the middle of a paragraph. However, an illustration requiring frequent reference from multiple locations of the document may be given at the end of the document.

A.2.13.2 Line Graphs

Line Graphs are best used to depict trends or changes over time, or relationships between two or more variables. The guidelines for preparing Line Graphs are:

1. The independent variable shall be represented on the **X**-axis
2. The dependent variable shall be represented on the **Y**-axis
3. Keep the plot lines to the minimum—in any case not more than five per graph
4. When more than one plot line is used, make the lines clearly distinguishable by the use of colour, solid lines, dotted lines, dashes etc.
5. Include a legend within the graph
6. Begin the **Y**-axis at zero. **X**-axis need not begin at zero
7. Keep all the gradations equal
8. Label items on the axes.

A.2.13.3 Bar Charts

Bar Charts are best used to compare multiple alternatives with each alternative depicted as a bar. The guidelines for preparing Bar Charts are:

1. Begin **Y**-axis at zero.
2. Break the bar with wavy lines or slash marks between zero and the lowest value when the bar is too long.
3. Keep all the vertical gradations equal and all the horizontal gradations equal.
4. Keep width of all bars equal. Keep the width of the space between bars equal. Make the bars wider than the space between the bars.
5. Arrange the bars in some order—alphabetically, chronologically, ascending or descending order etc.
6. Distinguish between the bars by using contrasting colours/hatch-patterns.
7. Include a legend within the graph.
8. Label items on the axes and the bars. Place figures within the bars or at the top of each bar to give accurate values.

A.2.13.4 Pie Charts

Pie Charts are best used to depict apportionment of a single entity on various heads. The pie or the circle is the whole; the slices or segments are the apportionments. The parts must add up to 100 %. The guidelines for preparing Bar Charts are:

1. There shall be at least three segments—otherwise, the chart is unnecessary.
2. Restrict the number of segments to seven. This can be achieved by clubbing minor segments into one single "Miscellaneous" or "Others" segment. This segment can be explained in a footnote.
3. Identify each segment either within the circle or outside the circle using a guideline.
4. Distinguish between the segments using contrasting colors or hatch patterns
5. Make the size of the pie appropriate for the page.

A.3 Language

It is not possible to include all the rules of English grammar in this guideline. However, a few guidelines are included here to achieve uniformity and to assist as a reference guide for writing Correct English.

A.3.1 Sentence Construction

While it is possible to construct sentences with one or two words, it is recommended that the sentences be constructed with at least three words. The characteristics of a good sentence are:

1. It is short, simple and clear
2. It has a single central idea
3. It has agreement between its parts
4. Verbs are properly used
5. Pronouns are properly used
6. It is preferably affirmative
7. It is preferably in active voice.

A.3.1.1 Short, Simple and Clear

To keep sentences short, simple and clear are:

1. Use simple and familiar words
2. Avoid unnecessary words
3. Use simple sentences as far as possible

4. When writing complex sentences, restrict sub-ordinate clauses to two
5. When writing compound sentences restrict the number of conjunctions to two

A.3.1.2 Single Central Idea

This can be achieved by focussing on the subject and by not introducing additional subjects in the sentence.

A.3.1.3 Agreement Between Parts

There must be consistency and agreement between parts of the sentence:

1. Subject and verb must agree with each other.
2. Number—Singular/Plural—should be consistent through the sentence—correct examples are:

   a. A **list** of guidelines **was** given to the auditor.
   b. **Mr. Smith**, as well as his two assistants, **was** working on the assignment.
   c. The **software processes**, which are maintained by QA, **are** extremely complex.

3. Words linked to the subject by expressions such as 'together with', 'as well as', 'along with', 'including', 'and not', 'in addition to', do not affect the number of the verb. The correct examples are:

   a. The **Project Leader**, as well as his two Module Leaders and five Team Members, **writes** efficient code.
   b. The **Vice President**, and not the Business Unit Heads, **was** charged with the responsibility.

4. When the subject is any of the following words or is modified by them, the verb must be singular. Each, Everybody, Anybody, Nobody, Every, A person, Either. Example—Each one of the group has a responsibility.
5. When the subject is a collective noun, the meaning to be conveyed shall determine whether the subject is singular or plural. Example—The **committee were** equally divided in supporting the two proposals.

A.3.2   Proper Usage of Verbs

Tenses of verbs in a sentence will accurately indicate the correct sequence of actions. The verb in a sub-ordinate clause shall therefore take a tense consistent with the verb in the main clause. Here are some examples of some correct examples:

1. When the machine **stopped**, the foreman realized that no one **had oiled** it.
2. When he had come, I left.

3. When he came, I left.
4. Before he came, I left.
5. Because he came, I left.

When a participle is used in a phrase, there must be something for the phrase to modify, to cling to or depend upon. Consider this sentence—"While watching TV, an argument broke out." What is wrong in the above sentence? An argument cannot watch TV.

Modifiers must be located so that it is clear what they modify. Consider this sentence "Even though it will take six years for the machines to pay for themselves, *if conditions do not bring about a change in prices*, the investment is decidedly attractive in the long run." Does the italicized clause in the above sentence refer to "*it will take six years ...*" or "to "*the investment is ...*"

A.3.3 Proper Usage of Pronouns

Pronouns must refer unmistakably to their antecedents, such as who, which, that, must be placed as close to their antecedents as possible.

Consider this sentence "We are sending you a check for the defective part, which we hope would be satisfactory." To what does "*which*" refer to? To "*part*" or to "*check*"?

Consider another sentence "*She* had already informed the *typist* that *she* would be responsible for the general form of the letters." Who would be responsible—"*typist*" or "*she*:?

A.3.4 Punctuation

Use commas to set off non-restrictive clauses, introduced usually by such words as who, which, that and where. Here are some correct examples:

1. A refrigerator, *which is a necessity to American housewives*, is a luxury in most parts of the world.
2. Mr. Johnson, who has been with us many years, has earned an enviable reputation in our department.
3. Washington, where the White House is located, is the capital of USA.

Use a semi-colon to separate two independent clauses not connected by a coordinating conjunction. Here are some correct examples:

1. We shall send your merchandize on September 25th; this should arrive in ample time for your Christmas sale.
2. The report was submitted on time; the resulting action corrected the difficulty.

Use of colon is not recommended where the practice of using soft copies is prevalent as colon would not be clearly visible. Usage of hyphen/dash (–) is recommended to introduce a formal list. Here is an example.

1. There are three steps in this procedure −

    a. Analyze the job carefully
    b. Eliminate unnecessary details
    c. Reduce operations to routine wherever possible.

The following are the guideline in using the "Apostrophe":

1. Use apostrophe to indicate possession:

    a. For nouns not ending in 's', place an apostrophe followed by an 's'. Example—The client's response is expected tomorrow.

2. For nouns ending in 's', place an apostrophe after the noun. Example—EDS' strength is in maintaining large applications.
3. Use an apostrophe to show that letters have been omitted. Examples—Don't, could've, it's.

A.3.5 Numbers

Here are the guidelines for handling numbers inside the text:

1. Write out numbers below ten and numbers divisible by ten up to hundred.
2. If a sentence begins with a number, write it.
3. When numbers are expressed in words, use a hyphen to join compound numbers, such as twenty-three, sixty-one.
4. When a sentence begins with a number, followed by another number to form an approximation, both should be expressed in words. Example—"Twenty or Twenty-five days will be needed to finish the task."
5. When a sentence contains two series of numbers, the number in one series should be expressed in words and the other series should be expressed in figures. Example—"Five students scored 95 %; seventeen students scored 80 %; and eleven scored 75 %."
6. When one number immediately precedes another number of a different context, one number should be expressed in words and the other in figures. Examples:

    a. We ordered twenty-five 10 by 12 prints.
    b. The specification calls for four teams of 12 persons each.

7. Amounts of money, generally, should be expressed in figures.
8. The following should be expressed in figures, generally.

    a. Dates
    b. Street numbers
    c. Numbered items (page xx, Chap. 2 etc.)
    d. Decimals

    e. Dimensions

    f. Percentages

    g. Fractions

9. A zero shall appear before the decimal point if it is not preceded by a numeral.

10. Common decimals less than one shall be used in the singular but any number greater than one shall be used in the plural.

A.3.6 Capitalization

Here are the guidelines in capitalization of letters:

1. Capitalize the first word and all other principal words in titles of books, reports or business documents. Examples:

   a. Software Requirements Document

   b. Analysis of the Tool for Project Management

2. Capitalize all the words of names used to identify organizations, places, buildings etc. Examples:

   a. Cyber Towers

   b. International Business Machines Inc.

   c. New York

3. Capitalize the first word of the following:

   a. Complete sentences

   b. Bullets

   c. Quotations, only if the quotation is a complete statement. Example—Your letter said "Send the replacement by air."

4. Do not capitalize words in the following:

a. General terms which do not identify a specific person, place or thing. Examples—

   i. "a doctor (but Dr. Smith)

   ii. our president (but President Regan)

   iii. my uncle (but Uncle Rockefeller)

b. The names of seasons. Example—spring, summer, winter

c. Nouns used with numbers. Examples:

   i. page 57

   ii. type 1

   iii. method 3

A.3.7   Usage of Words

Here are the guidelines for using articles:

A.3.7.1   Articles

AN—The rule governing the use of article 'an' is phonetic and not orthogonal. The article 'An' is used before vowels and a silent 'h'. Examples:

1. an eyelet
2. an 18th century practice
3. an heirloom"

A—The article 'a' is used before all consonants and also before vowels preceded by the sound 'y' or 'w'. Examples:

1. a unit
2. a one-room
3. a eulogy

THE—The definite article 'the' is applied to an individual object or objects mentioned earlier in the text or already known, or contextually particularized. When in doubt, the answer to 'what' or 'which' generally clarifies whether 'the' is required or not. Examples:

1. The software developed at Microsoft is commercial in nature. Which software? 'The' software developed at Microsoft.

The definite article 'the' should not be used when objects are referred to in general or by an undefined sense. Example:

1. India, in comparison with other countries of the world, ought to have 20 million cars.

    a. Which countries? Other countries in general.

A.3.7.2   AND/OR

The use of the expression 'and/or' is not recommended except in tables. Examples:

1. NEMA and/or IEEE abbreviations may be used—NOT RECOMMENDED
2. NEMA abbreviations or IEEE abbreviations may be used—RECOMMENDED

A.3.7.3   Usage of 'etc.'

Use of 'etc.' shall be avoided as far as possible. If it is found absolutely necessary to use "etc.", it should be used after three or more nouns without the conjunction 'and'. In particular, 'etc.' shall not be used after a sequence introduced by expressions like 'for example' and 'such as'.

A.3.7.4 Shall, Should, Will, Would, Must, May and Can

1. Shall—The word 'shall' will be used to indicate the advisory (not obligatory) character of a requirement.
2. Should—The word 'should' will be used to indicate that the requirement is advisory and not obligatory.
3. Will—The word 'will' will be used to indicate the obligatory (not advisory) character of a requirement.
4. Would—The word 'would' will be used to indicate that the requirement is obligatory (not advisory).
5. Must—The word 'must' will not be used to express obligatory character.
6. May—The word 'may' will be used when permissive character or probability of occurrence is implied.
7. Can—The word 'can' will not be used to express either permissive character or probability of an occurrence.

A.4 Recommended and Not Recommended Usage

It may be noted that the list given below is by no means complete. However, this is a first list that may be improved further.

Not recommended	Recommended
At the present time	Now
Despite the fact that	Although
For the reason that	Since/because
On behalf of	For
So as to	To
Inter alia	Among other things
Mutatis mutandis	With necessary alterations
Ipso facto	By the very fact
Erroneous	Wrong
Annihilate	Destroy
Recapitulate	Recall
Facilitate	Help
Utilize	Use
Competence	Skill
Above mentioned clauses	Clauses mentioned above
Adequate enough	Adequate or Enough
Afore mentioned	Mentioned earlier
Brief details	Details
Enclosed herewith	Enclosed
Equipments	Equipment
May invariably	Shall invariably

(continued)

(continued)

Not recommended	Recommended
Often times	Often
Personnels	Personnel
Sufficient enough	Sufficient or Enough
Two times	Twice
As per	'In conformance' with or ' according to' or 'adhering to' or 'in adherence with'
Kindly	Please
Advance planning	Planning
As to whether	Whether
Basic fundamentals	Fundamentals
But nevertheless	Nevertheless
Close scrutiny	Scrutiny
Collaborate together	Collaborate
Complete monopoly	Monopoly
Definite Decision	Decision
Difficult dilemma	Dilemma
Descend down	Descend
Direct confrontation	Confrontation
During the course of	During
Eradicate completely	Eradicate
Estimated at about	Estimated at
Estimated roughly at	Estimated at
Every now and then	Now and then
Exact opposites	Opposites
Few in number	Few
Free gift	Gift
Future plans	Plans
Integral part of	Part of
Invited guests	Guests
Just recently	Recently
Midway between	Midway or between
Might possibly	Might or possibly
My personal opinion	My opinion
Over exaggerate	Exaggerate
Period of time	Period
Plan ahead	Plan
Refer back	Refer
Repeat again	Repeat
Reply back	Reply
Reported to the effect that	Reported that
Revert back	Revert
Spell out in detail	Spell out
Sum total	Sum or Total
Sworn affidavits	Affidavits
True facts	Facts

A.5 Final words of Documentation Guidelines

This guideline assumes that the reader is competent in basic grammar of English language. I am giving the finer points of documenting requirements documents correctly and facilitate identical interpretation of information presented. When we document requirements in software engineering, one of the main objectives is to avoid the possibility for confusion. We need to ensure that the reader would arrive at the same interpretation as the writer intended. These guidelines are a step in that direction. Every professional standards organization would have a documentation guideline and ensure that all the documents released by them are conforming to that guideline. Software organizations also produce a significant amount of documentation that would be referred to by individuals other than the author. Therefore, it is essential for software development organizations to have a documentation guideline. The guidelines presented here may be used as they are or with modification.

Appendix B
Planguage

B.1 Introduction

Planguage was developed over a period of time by Tom Gilb (www.gilb.com) during his work as a faculty member, as well as, a consultant to various professional organizations such as HP, IBM, Intel, Philips, Ericson and so on. It was intended for use in systems engineering to improve the quality of specifications. It was published in book form in 2005 under the title of "Competitive Engineering: A Hand Book for Requirements Engineering, and Software Engineering using Planguage", authored by Tom Gilb and published by Elsevier Butterworth-Heinemann, USA. This appendix is an excerpt from that book with permission from Tom Gilb. I record here my sincere thanks to Mr. Tom Gilb for giving me permission to include details about Planguage in this book. If any reader wishes to have a comprehensive understanding of Planguage, I suggest reading the above mentioned book which is about 500 pages of tightly-packed information.

Planguage gives us tools to tackle large and complex systems with an objective of reducing the risk of waste, delay and failure. Planguage should be viewed as a powerful way to develop and implement strategies that will help the project to deliver the required competitive results. Planguage focuses on reducing or eliminating the ambiguity in the specifications.

B.2 Planguage

Planguage consists of a specification language and a corresponding set of process descriptions. These two (Planguage terms and processes) are used together. The Planguage terms consist of concepts, grammar, and icons. The process descriptions consist of best practices for carrying out certain tasks. These are, Requirement Specification, Design Engineering, Specification Quality Control, Impact Estimation,

M. Chemuturi, *Requirements Engineering and Management for Software Development Projects*, DOI: 10.1007/978-1-4614-5377-2,
© Springer Science+Business Media New York 2013

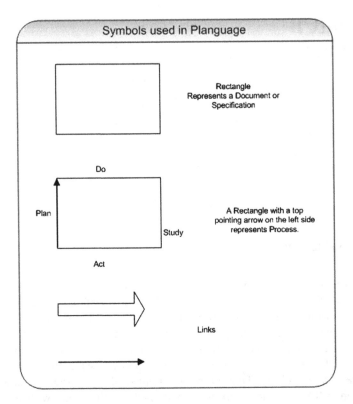

Fig. B.1 Symbols used in planguage diagrams

and Evolutionary Project Management. The symbols used in Planguage diagrams are depicted in Fig. B.1.

Tom Gilb suggests a template for capturing the project requirements and preparing the Requirement Specification (Tom Gilb was very specific in his book to use the term "Requirement" in place of the commonly used "Requirements". This may be noted.) It is depicted in Table B.1.

Function Specification Template is depicted in Table B.2.

Scalar Requirement Template is depicted in Table B.3.

Table B.1 Requirement Specification Template

Requirement Specification Template (A Summary Template)

Tag: <Tag Name for the System>
Type: System

==================== Basic Information ==================

Version: <Date or other version number>
Status: <Draft / SQC / Exited / Approved / Rejected>
Quality Level: <Maximum remaining major defects/page, sample size, date>
Owner: <Role/email/name of the person responsible for changes and updates>
Stakeholders: <Name any stakeholders (other than the owner) with an interest in the system>
Gist: <A brief description of the system>
Description: <A full description of the system>
Vision: <The overall aims and direction for the system>

==================== Relationships ====================

Consists of: Subsystem: <Tags for the immediate hierarchical subsystems if any, comprising this system>
Linked to: <Other systems or program that this system interfaces with>

=============== Function Requirements =================

Mission: <Mission statement or tag of the mission statement>
Function Requirement:
<{**Function Target, Function Constraint**}>: <State tags of the function requirements>
Note: 1. See Function Specification Template
        2. By default, 'Function Requirement' means 'Function Target'

============= Performance Requirements ================

Performance Requirement:
<{**Quality, Resource Saving, Workload Capacity**}>: <State tags of the performance requirements>
Note: See Scalar Requirement Template

(continued)

Table B.1 (continued)

================ **Resource Requirements** ===================

Resource Requirement:
<{Financial Resource, Time Resource, Headcount Resource, others}>: <State
tags of the resource requirements>
Note : See Scalar Requirement Template

==================== **Design Constraints** ====================

Design Constraint: <State tags of any relevant design constraints>

================ **Condition Constraints** ====================

Condition Constraint: <State tags of any relevant condition constraints or specify a list of condition constraints>

============== **Priority and Risk Management** ================

Rationale: <What are the reasons supporting these requirements?>
Value: <State overall stakeholder value associated with these requirements>
Assumptions: <Any assumptions that have been made>
Dependencies: <Using text or tags, name any major system dependencies>
Risks: <List or refer to tags of any major risks that could cause delay or negative
impacts to the achieving of the requirements>
Priority: <Are there any known overall priority requirements?>
Issues: <Unresolved concerns or problems in the specification or the system>

========= **Evolutionary Project Management Plan** =============

Evo Plan: <State the tag of the Evolutionary Project Management Plan>

================= **Potential Design Ideas** ==================

Design Ideas: <State tags of any design ideas for this system, which are not in
the Evo Plan>

Table B.2 Function Specification Template

Function Specification Template

\<Function Tag1\>

Type: Function

Description: \<Describe the function here, well enough to allow testing of it\>

Attribute 1: Scale \<?\> Goal: \<?\>
Attribute 2: Scale \<?\> Goal: \<?\>
Attribute n: Scale \<?\> Goal: \<?\>

Note :

  1. Scale is the precision of the attribute
  2. Goal is the usage of the attribute

Table B.3 Scalar Requirement Template

Elementary Scalar Requirement Template

Tag: \<Tag name of the elementary scalar requirement\>
Type:
\<{Performance Requirement: {Quality Requirement, Resource Saving Requirement, Workload Capacity Requirement}. Resource Requirement: {Financial Requirement, Time Requirement, Headcount Requirement, others }}\>

=================== **Basic Information** ===================

Version: \<Date or other version number\>
Status: \<Draft / SQC / Exited / Approved, / Rejected\>
Quality Level: \<Maximum remaining major defects/page, sample size, date\>
Owner: \<Role/email/name of the person responsible for changes and updates\>
Stakeholders: \<Name any stakeholders (other than the owner) with an interest in the system\>
Gist: \<A brief description, capturing the essential meaning of the requirement \>
Description: \<Optional, full description of the requirement\>
Ambition: \<Summarize the ambition level of only the targets below. Give the overall real ambition level in 5 – 20 words.\>

=================== **Scale of Measure** ===================

Scale: \<Scale of measure for the requirement (States the units of measure for all the targets, constraints and benchmarks) and the scale qualifiers\>

(continued)

=================== **Measurement** ===================

Meter: <The method to be used to obtain measurements on the defined Scale>

======== **Benchmarks** ====== **Past Numeric Values** ==========

Past [<when, where, if]: <Past or current level. State if it is an estimate> <-
<Source>.
Record [<when, where, if]: <State-of-the-art level> <-<Source>
Trend [<when, where, if]: <Prediction of the rate of the change or future state-of-
the-art level> <-<Source>

======== **Targets** ========== **Future Numeric Values** ========

Goal/Budget [<when, where, if]: <Planned target level> <-<Source>
Stretch [<when, where, if]: <Motivating ambition level> <-<Source>
Wish [<when, where, if]: <Dream level (unbudgeted> <-<Source>

======== **Constraints** =========== **Specific Restrictions** =======

Fail [<when, where, if]: <Failure level> <-<Source>
Survival [<when, where, if]: <Survival level> <-<Source>

=================== **Relationships** ===================

Is part of: <Refer to the tags of any supra-requirements (complex requirements)
that this requirement is part of. A hierarchy of tags (For example A, B, C) is pre-
ferable>
Is impacted by: <Refer to the tags of any design ideas that impact this require-
ment> <-<Source>
Impacts: <name any requirements or designs or plans that are impacted by this
requirement>

============ **Priority and Risk Management** ================

Rationale: <Justify why this requirement exists>
Value: <Name [Stakeholder, time, place, event]: Quality, or express in words, the
value claimed as a result of delivering the requirement >
Assumptions: <State any assumptions made in connection with this requirement>
<-<Source>
Dependencies: <State anything that achieving the planned requirement level is
dependent on> <-<Source>
Risks: <List or refer to tags of any major risks that could cause delay or negative
impact > <-<Source>
Priority: <List the tags of any system elements that must be implemented before
or after this requirement>
Issues: <State any known issues>

B.3 Glossary of Planguage Terms

Now, I present you the glossary of the terms used in Planguage. I will be covering all the terms used for requirements engineering activity and leaving out the rest. Also, I would not be covering the entire description given in the book here as it will be repetition. The idea is to introduce you to the Planguage and if you are interested in learning more, I suggest reading the book by Tom Gilb mentioned earlier.

Planguage is a full-fledged method for carrying out software engineering. It needs significant effort and little time from you to master it and use it expertly. If you put in the effort and spend the time, I am sure you can reap rich benefits from using Planguage.

Term	Explanation
A	
After	Indicates a planned sequencing of events
Aim	A stated desire to achieve something by certain stakeholders
Ambition	A parameter which can be used to summarize the ambition level of a performance or resource target requirement
And	A logical operator to join any two expressions
Assumption	Unproven conditions, which if not true at some defined point in time, would threaten something, such as the volatility of specification or the achievement of our requirements
Attribute	An observable characteristic of a system.
Author	A person, who writes or updates a document or specification of any kind
Authority	A specific level of power to 'decide' or 'influence' a specific matter requiring some degree of judgment or evaluation
B	
Background	Background information is the part of a specification, which is useful related information, but is not central (core) to the implementation nor is it commentary
Backroom	Refers to a conceptual place used to describe any process or activities that are not necessarily visible upfront
Baseline	A set of system attribute specifications that defines the state of a given system
Basis	An underlying idea that is a foundation for a specification
Before	A parameter used to indicate planned sequencing of events
Benchmark	A specified reference point or baseline
Benefit	Value delivered to stakeholders
Binary	An adjective used to describe objects which are specified as observable in two states
C	
Catastrophe	Level of an attribute where disaster threatens all or a part of a system
Checklist	A list of questions used to check a document for completeness or accuracy
Complex	A complex component is composed of more than one elementary and/or complex component

(continued)

(continued)

Term	Explanation
Condition	A specified pre-requisite for making a specification or a system component valid
Condition Constraint	A requirement that imposes a conscious restriction for a specified system scope
Consists of	A parameter used to list a complete set of the sub-components or elements comprising a component
Constraint	A requirement that explicitly and intentionally tries to directly restrict any system or process
Core Specification	It will result in real system changes being made; incorrect core specification would materially and negatively affect the system in terms of costs, effort or quality.
Cost	An expense incurred in building or maintaining a system
Credibility	The strength of belief in and hence validity of information
Critical Factor	A scalar attribute level, a binary attribute or condition in a system, which can on its own, determine the success or failure of the system under specified conditions
D	
Definition	A parameter that is used to define a tagged item
Dependency	A reliance of some kind of one set of components on another set of components
Description	A set of words and/or diagrams, which describe and partially define a component
Deviation	The amount of value (estimated or actual) by which some attribute differs from some specific benchmark or target
Due	A parameter indicating when some aspect of a specification is due
During	Used when specifying events to indicate a time dependency for events that must be carried out concurrently
E	
Elementary	Elementary component is not decomposed into sub-components
Error	Something done incorrectly by a human being
Estimate	A numeric judgment about a future, present or past level a scalar system attribute
Event	A specified occurrence
Evidence	The historical facts, which support an assertion.
Evolutionary	It implies association with an iterative process of change, feedback, learning and consequent change
Except	Used to specify that the following term or expression is an exception from the previous term or expression
F	
Fail	Signals an undesirable and unacceptable system state
Frontroom	It refers to a conceptual place used to describe any process or activities
Function	It is what a system does
Function Constraint	A requirement which places a restriction on the functionality that may exist in a system
Function Requirement	Specifies that the presence or absence of a defined function is required

(continued)

(continued)

Term	Explanation
Function Target	A specified function requirement. We need to plan delivery of the function under the specified conditions
Fuzzy	A specification which is known to be somewhat unclear, potentially incorrect or incomplete
G	
Gap	For a scalar attribute gap is a range from an impact estimate or a specification benchmark to a specification target
Gist	A parameter used to state the essence, or main point, of a specification
Goal	A primary numeric target level of performance
I	
Icon	In Planguage an icon is a symbol that is keyed (keyed icon) or drawn (drawn icon) that represents a concept
If	A logical operator used in qualifiers to explicitly specify conditions
Impact	The estimated or actual numeric effect on a requirement attribute under given conditions
Impacts	A parameter that is used to identify the set of attributes that are considered likely to be impacted by a given attribute
Includes	Expresses the concept of inclusion of a set of components within larger set of components
Issue	Any subject of concern that needs to be noted for analysis and resolution
L	
Level	A defined numeric position on a scale of measure
Limit	A numerical level at a border, that is, at an edge of a scalar range
M	
Master Definition	The primary and authoritative source of information about the meaning of a specification or a specification element
Metric	Any kind of numerically expressed system attribute
Mission	A mission specifies who we are (or what we do) in relation to the rest of the world
O	
Objective	A synonym of performance requirement
Or	A logical operator used in qualifiers or other appropriate specifications
Or Better	An expression used within a scalar specification to explicitly emphasize that the specified level has a range of acceptable values rather than being just a fixed single value
Or Worse	An expression used within a scalar specification to explicitly emphasize that the specified level has a range of acceptable values rather than being just a fixed single value
Owner	A person or group responsible for an object and for authorizing any change to it
P	
Parameter	A Planguage defined term. Parameters are always written with at least a leading capital letter to signal the existence of a formal definition
Past	A parameter used to specify historical reference, a benchmark.
Percentage Uncertainty	It is calculated from the scale uncertainty baseline and target data

(continued)

(continued)

Term	Explanation
Performance	An attribute set that describes measurably 'how good' the system is at delivering effectiveness to its stakeholders
Performance Constraint	Specifies some upper and lower limits for an elementary scalar performance attribute
Performance Requirement	Specifies the stakeholder requirements for 'how well' a system should perform
Performance Target	Is a stakeholder-valued numeric level of system performance
Place	Defines where
Planguage	Tom Gilb © A specification language and a set of related methods for systems engineering
Priority	The determination of a relative claim on limited resources
Procedure	A repeatable description to instruct people as to the best-known practice, or recommended way, to carry out the task of a defined process
Process	Is a work activity which consists of an entry process, entry conditions, a task process, and an exit process
Q	
Qualifier	A defined set of conditions embedded in, or referenced by, a specification
Quality	A scalar attribute reflecting 'how well' a system functions
Quality Level	A measure of a specification's conformance to any specified relevant standards
R	
Range	The extent between and including two defined numeric levels on a scale of measure
Rationale	The reasoning or a principle that explains and thus seeks to justify a specification
Readership	The readership of a specification is all the 'types of people' we intend shall read or use the specification
Record	A parameter used to inform us about an interesting extreme of achievement
Relationship	A connection between two objects
Requirement	A stakeholder-desired, or needed, target or constraint
Requirement Engineering	A requirement process carried out with an engineering level of rigor
Requirement Specification	A defined set of requirements
Resource	Any potential system input (time, money, effort, space, data and any other)
Resource Constraint	A resource requirement, which specifically restricts or serves as a warning about the level that can be used of a resource
Resource Requirement	Specifies how much of a resource should be made available for later consumption
Resource Saving	A performance attribute of a system
Resource Target	It is a budget.
Review	Any process of human examination of ideas with a defined purpose and a defined standards of inquiry
Risk	Any factor that could result in a future negative consequence
Role	A defined responsibility, interest, or scope for people

(continued)

(continued)

Term	Explanation
Rule	Any statement of a standard on how to write or carry out some part of a systems engineering or business process
S	
Safety Deviation	A measure of the estimated-or-observed difference between a required safety margin and the estimated or actual system attribute level
Safety Factor	The dimensionless ratio of 'conscious over-design' that is either required or actually applied to some part of the system
Safety Margin	A scalar difference between a required defined target or constraint level, and its calculated safety level derived using the appropriate safety factor
Scalar	An adjective used to describe objects, which passes or are measured using at least one scale of measure
Scale	A parameter used to define a scale of measure
Scale Impact	Is an absolute numeric value on the scale of measure
Scale Uncertainty	An estimate of the error margins for a specific scale impact
Scope	Describes the extent of influence of something
Software Engineering	The discipline of making software systems deliver the required value to all stakeholders
Source	A synonym for process input information
Specification	Communicates one or more system ideas and/or descriptions to an intended audience
Stakeholder	Any person, group or object, which has some direct or indirect interest in a system
Standards	An official, written specification that guides a defined group of people in doing a process. It is a best-known practice
Status	The outcome of an evolution of a defined condition
Stretch	A parameter used to define a somewhat more ambitious target level than the committed goal or budget levels
Supports	Used to indicate what an attribute is mainly intended to support
Survival	A state where the system can exist
System (Planguage)	Any useful subset of the universe that we choose to specify. It can be conceptual or real. In Planguage a system can be described fundamentally by a set of attributes
System Engineering	An engineering process encompassing and managing all relevant system stakeholder requirements, as well as all design solutions and necessary technology, economic and political areas
T	
Tag	A term that serves to identify a statement, or set of statements, unambiguously
Target	A specified stakeholder-valued requirement which you are aiming to deliver under specified conditions
Task	A defined and limited piece of work
Test	To plan and execute an analytical process on any system, product or process, where we attempt to understand if the system performs as expected or not

(continued)

(continued)

Term	Explanation
Time	Defines 'when'
Trend	A parameter used to specify how we expect or estimate attribute levels to be in the future
Type	Specifies the category of a Planguage concept
Uncertainty	The degree to which we are in doubt about how an impact estimate, or measurement, of an attribute reflects reality
U	
Until	A logical operator that is used to limit the extent of a scalar range of values
User-defined Term	A definition of a term made by a user
V	
Value	Perceived benefit: that is, the benefit we think we will get from something
Version	An initial or changed specification instance
Vision	An idea about a future state, which is very long range and probably idealistic, may be even unrealistic
W	
Wish	A parameter used to specify a stakeholder-valued, uncommitted target level for a scalar attribute
Workload Capacity	A performance attribute. It is used to express the capacity of a system to carry out its workload, that is, 'how much' a system can do, did or will do

About the Author

Murali Chemuturi is an information technology and software development subject matter expert, hands-on pro-grammer, author, consultant and trainer. Since 2001, he is offering consultancy on information technology and train-ing to organizations in India and in the USA from Chemuturi Consultants.

Chemuturi Consultants also offers a number of products to aid project managers and software development professionals such as PMPal, a software project management tool; and Estimator-Pal, FPAPal and UCPPal, a set of software

estimation tools. Chemuturi Consultants also offers a material requirements planning software product MRPPal to assist small to medium manufacturing organizations to efficiently manage their materials.

Prior to starting his own firm, Murali gained over 15 years of industrial experience in various engineering and manufacturing management positions. He then gained more than 26 years of information technology and software development experience. His most recent position prior to forming his firm was Vice President of Software Development at Vistaar e-Business Pvt., Ltd.

Mr. Chemuturi's undergraduate degrees and diplomas are in Electrical and Industrial Engineering and he holds a MBA and a Post Graduate Diploma in Computer Methods and Programming. He has several years of academic experience teaching a variety of computer and IT courses such as COBOL, Fortran, BASIC, Computer Architecture, and Database Management Systems.

M. Chemuturi, *Requirements Engineering and Management for Software Development Projects*, DOI: 10.1007/978-1-4614-5377-2, © Springer Science+Business Media New York 2013

Mr. Chemuturi authored two books, namely "Software Estimation: Best Practices, Tools and Techniques for Software Project Estimators" and "Mastering Software Quality Assurance: Best Practices, Tools and Techniques for Software Developers" published in the USA by J. Ross Publishing, Inc, He co-authored another book with Thomas M. Cagley, Jr. titled "Mastering Software Project Management: Best Practices, Tools and Techniques", published by J. Ross Publishing, Inc of the USA.

Murali is a senior member of IEEE, a senior member of the Computer Society of India and a Fellow of the Indian Institute of Industrial Engineering and he is a well published author in professional journals.

Index

M. Chemuturi, *Requirements Engineering and Management for Software*
Development Projects, DOI: 10.1007/978-1-4614-5377-2,
© Springer Science+Business Media New York 2013